U.S. Department of
Transportation

DOT-VNTSC-NASA-12-01
NASA/TM-2012-217783

Behind Start of Take-off Roll Aircraft Sound Level Directivity Study

Final Report
December 2012

Prepared for:
National Aeronautics and Space Administration
Langley Research Center
Structural Acoustics Branch Division
Hampton, VA 23681-2199

Prepared by:
U.S. Department of Transportation
Research and Innovative Technology Administration
John A. Volpe National Transportation Systems Center
Environmental Measurement and Modeling Division
Cambridge, MA 02142

Notice

This document is disseminated under the sponsorship of the Department of Transportation in the interest of information exchange. The United States Government assumes no liability for its contents or use thereof.

Notice

The United States Government does not endorse products or manufacturers. Trade or manufacturers' names appear herein solely because they are considered essential to the objective of this report.

REPORT DOCUMENTATION PAGE			Form Approved OMB No. 0704-0188
Public reporting burden for this collection of information is estimated to average 1 hour per response, including the time for reviewing instructions, searching existing data sources, gathering and maintaining the data needed, and completing and reviewing the collection of information. Send comments regarding this burden estimate or any other aspect of this collection of information, including suggestions for reducing this burden, to Washington Headquarters Services, Directorate for Information Operations and Reports, 1215 Jefferson Davis Highway, Suite 1204, Arlington, VA 22202-4302, and to the Office of Management and Budget, Paperwork Reduction Project (0704-0188), Washington, DC 20503.			
1. AGENCY USE ONLY (Leave blank)	2. REPORT DATE December 2012	3. REPORT TYPE AND DATES COVERED Final Report	
4. TITLE AND SUBTITLE Behind Start of Take-off Roll Aircraft Sound Level Directivity Study		5. FUNDING NUMBERS VX83 - BS243, FA4SCD LTA59	
6. AUTHOR(S) Michael C. Lau, Christopher J. Roof, Gregg G. Fleming, Amanda S. Rapoza, Eric R. Boeker, David A. McCurdy, and Kevin P. Shepherd			
7. PERFORMING ORGANIZATION NAME(S) AND ADDRESS(ES) U.S. Department of Transportation Research and Innovative Technology Administration John A. Volpe National Transportation Systems Center Environmental Measurement and Modeling Division, RVT-41 Cambridge, MA 02142-1093		8. PERFORMING ORGANIZATION REPORT NUMBER DOT-VNTSC-NASA-12-01	
9. SPONSORING/MONITORING AGENCY NAME(S) AND ADDRESS(ES) National Aeronautics and Space Administration Langley Research Center, Structural Acoustics Branch Mail Stop 463, Hampton, VA 23681-2199		10. SPONSORING/MONITORING AGENCY REPORT NUMBER NASA/TM-2012-217783	
11. SUPPLEMENTARY NOTES NASA Program Managers: Kevin P. Shepherd, David A. McCurdy			
12a. DISTRIBUTION/AVAILABILITY STATEMENT		12b. DISTRIBUTION CODE	
13. ABSTRACT (Maximum 200 words) The National Aeronautics and Space Administration (NASA), Langley Research Center (LaRC) and the Environmental Measurement and Modeling Division of the United States Department of Transportation's John A. Volpe National Transportation Systems Center (Volpe) conducted a noise measurement study to examine the sound level directivity pattern behind aircraft start-of-takeoff roll. This report discusses the procedures and methodologies used to measure and quantify data acquired from the Study.			
14. SUBJECT TERMS noise, aircraft noise, airport noise, behind start of take-off roll, aircraft directivity patterns, take-off roll noise, INM, Integrated Noise Model, AEDT, Aviation Environmental Design Tool, noise prediction, computer noise model		15. NUMBER OF PAGES 89	
		16. PRICE CODE	
17. SECURITY CLASSIFICATION OF REPORT Unclassified	18. SECURITY CLASSIFICATION OF THIS PAGE Unclassified	19. SECURITY CLASSIFICATION OF ABSTRACT Unclassified	20. LIMITATION OF ABSTRACT

NSN 7540-01-280-5500

Standard Form 298 (Rev. 2-89)
Prescribed by ANSI Std. 239-18
298-102

TABLE OF CONTENTS

1 INTRODUCTION 1
   1.1 Background 1
   1.2 Objective 3
2 SITE SELECTION AND LOGISTICS 4
   2.1 Candidate Airport Selection.. 4
      2.1.1 Geographic Location... 4
      2.1.2 Airport Elevation 4
      2.1.3 Airport Fleet Mix 4
      2.1.4 Site Terrain, Near Runway 5
      2.1.5 Site Terrain, Distant 9
      2.1.6 Runway Configuration.. 10
      2.1.7 Other Factors Considered 10
      2.1.8 Site Selection Results.. 10
      2.1.9 Meteorological Data.. 10
      2.1.10 Selection of Study Airport 11
   2.2 Measurement Site Logistics 13
      2.2.1 Primary Runway for Measurements 13
      2.2.2 Aircraft Departure Schedule 13
      2.2.3 Preliminary Site Survey 16
3 INSTRUMENTATION 17
   3.1 Acoustic Instrumentation 17
      3.1.1 Specifications 18
      3.1.1.1 NASA Acoustic Vans 18
      3.1.1.2 Volpe Acoustic Measurement Systems 20
      3.1.2 Locations... 20
   3.2 Aircraft Tracking System.. 25
      3.2.1 Specifications.. 25
      3.2.2 Locations... 26
   3.3 Portable Weather Stations 26
      3.3.1 Specifications.. 26
      3.3.2 Locations... 27
   3.4 Tail Logging Instrumentation 27
      3.4.1 Specifications.. 27
      3.4.2 Locations... 28
   3.5 Other Instrumentation 28
      3.5.1 Specifications.. 29
      3.5.2 Locations... 29
4 MEASUREMENT PROCEDURES 30
   4.1 NASA Acoustic Vans 30
      4.1.1 Deployment... 30
      4.1.2 During Measurements... 30
   4.2 Volpe Acoustic Measurement Systems 31

    4.2.1 Deployment 31
    4.2.2 During Measurements 31
  4.3 Digital Video Tracking System 31
    4.3.1 Deployment 31
    4.3.2 During an Event 32
  4.4 Meteorological Station 32
    4.4.1 Deployment 32
    4.4.2 During an Event 32
  4.5 Aircraft Tail Number Loggers 32
    4.5.1 Deployment 32
    4.5.2 During an Event 32
  4.6 Test Director 33
    4.6.1 Deployment 33
    4.6.2 During an Event 33
  4.7 Communications 35
  4.8 Quality Assurance 35
5 DATA REDUCTION AND ANALYSIS 36
  5.1 Data Reduction Process 38
    5.1.1 Acoustic Data 38
    5.1.2 Aircraft Tail Numbers 39
    5.1.3 Meteorological Data 39
    5.1.4 Quality Assurance 40
      5.1.4.1 Take-off Roll Type 40
      5.1.4.2 Line-of-Sight Blockage 40
      5.1.4.3 Idling Aircraft and Terminal Noise 41
      5.1.4.4 Data Collected at 1950 ft 41
  5.2 Deriving Directivity Patterns 42
    5.2.1 Noise Metrics 42
    5.2.2 Normalizing Data to 285-Degrees 45
  5.3 Comparison of Newly Developed Patterns with AIR-1845 47
  5.4 Proposed Updated Directivity Adjustment 54
  5.5 Additional Directivity Investigations 57
  5.6 Case Studies 60
    5.6.1 Single Event Analysis 60
    5.6.2 Airport Analysis 64
6 CONCLUSIONS AND RECOMMENDATIONS 66
7 FUTURE RESEARCH 67
REFERENCES 68
APPENDIX A: SUMMARY OF INSTRUMENTATION LOCATIONS 69
APPENDIX B: TEST DIRECTOR LOG SHEET 70
APPENDIX C: TAIL NUMBER LOG SHEET 71
APPENDIX D: VOLPE NOISELOGGER SYSTEM SETUP 72
APPENDIX E: DIRECTIVITY PATTERNS AT 1950 FT 73
APPENDIX F: AIRCRAFT FLEET PERCENTAGE FOR DIRECTIVTY ADJUSTMENT
                DERIVATION 80

LIST OF TABLES

Table 1. Fleet Mix Rating Methodology... 5
Table 2. IAD Runway Usage Data 13
Table 3. Summary of Primary Instrumentation 17
Table 4. Totals for Recorded Departure Events... 36
Table 5. Totals for Turbo-Propeller Aircraft Events 36
Table 6. Totals for Aircraft in the "Other" Category.. 37
Table 7. Measured vs. INM-Modeled Comparison for a Microphone at 285-Degrees 47
Table 8. No. of Angular Offset Points which Exhibit Similarity between Aircraft..................... 58
Table 9. Horizontal Separation Distances between Engines 59
Table 10. Modeled Noise Levels behind Take-off roll Using Current SOTR Adjustment 61
Table 11. Modeled Noise Levels behind Take-off roll Using Proposed SOTR Adjustment 61
Table 12. Difference in Modeled Noise Levels for Two Different SOTR Implementations
(Current – Proposed).. 62
Table 13. Difference in Modeled Noise Levels for Two Different
SOTR Implementations (Current – Proposed) at Three Different Airports 64
Table 14. Summary of Instrument Locations 69
Table 15. Test Director Log Sheet... 70
Table 16. Tail Number Log Sheet.. 71
Table 17. Jet Aircraft Fleet Percentages for Directivity Adjustment Derivation 81

LIST OF FIGURES

Figure 1. Plots of Current Directivity Algorithm...2
Figure 2. Ideal Microphone Array 6
Figure 3. Aerial Photo of Runway 35L at Austin-Bergstrom Airport 7
Figure 4. An Aerial Photo of Runway 1R at Dulles International Airport 8
Figure 5. An Aerial Photo of Runway 9 at Tampa International Airport 8
Figure 6. Aerial Photo of Runway 17R at Austin-Bergstrom Airport 9
Figure 7. Airport Diagram of Dulles International Airport (IAD)... 12
Figure 8. Departure Flights at IAD During 10/1/03-10/7-03 14
Figure 9. Boeing 747 Departure Flights at IAD During 10/1/03-10/7-03 15
Figure 10. Airbus A340 Departure Flights at IAD During 10/1/03-10/7-03 15
Figure 11. NASA Acoustic Van 2, Designated as "Barth/72" 19
Figure 12. NASA Microphone System with Power Booster Box 19
Figure 13. Volpe Acoustic Measurement System... 20
Figure 14. Taxiway Turning Points Y, Z, Q on IAD Runway 30.. 21
Figure 15. Local Cartesian Coordinate System Used on Runway 30 21
Figure 16. Microphone Location Zones Behind Runway 30 22
Figure 17. Final Instrumentation Locations 24
Figure 18. Digital Video Tracking System 25
Figure 19. Transportable Automated Meteorological Station (TAMS)... 27
Figure 20. Primary Tail Number Logger 28
Figure 21. Differential Global Positioning System (dGPS) Receiver Antenna.............................. 29
Figure 22. Measurement Schedule... 30
Figure 23. Task Sequence for Each Event 34
Figure 24. NDAT Software Program... 38
Figure 25. Maximum Wind Speeds during Measurement Days.. 39
Figure 26. Symmetrical Microphones N1 and V1 at 900 ft... 41
Figure 27. Mean L_{AE} at 900 ft Microphone Locations for the Airbus A319........................... 42
Figure 28. Comparison between Different Metrics Used to Process A320 Events 44
Figure 29. Comparison between Different Metrics Used to Process B767 Events 44
Figure 30. AIR-1845 Directivity Pattern at 900 ft 45
Figure 31. Directivity Pattern for the Airbus A319 at 900 ft 46
Figure 32. Airbus A319 Directivity Pattern at 900 ft (16 Events)... 48
Figure 33. Airbus A320 Directivity Pattern at 900 ft (48 Events)... 48
Figure 34. Airbus A330 Directivity Pattern at 900 ft (4 Events)... 49
Figure 35. Airbus A340 Directivity Pattern at 900 ft (5 Events)... 49
Figure 36. Boeing 717 Directivity Pattern at 900 ft (3 Events)... 50
Figure 37. Boeing 737 Directivity Pattern at 900 ft (20 Events)... 50
Figure 38. Boeing 747 Directivity Pattern at 900 ft (7 Events)... 51
Figure 39. Boeing 757 Directivity Pattern at 900 ft (12 Events)... 51
Figure 40. Boeing 767 Directivity Pattern at 900 ft (27 Events)... 52
Figure 41. Boeing 777 Directivity Pattern at 900 ft (12 Events)... 52
Figure 42. McDonnell Douglas DC9 Directivity Pattern at 900 ft (23 Events) 53

Figure 43. Bombardier CL600 Directivity Pattern at 900 ft (104 Events) 53
Figure 44. Turbo-Propeller Aircraft Directivity Pattern at 900 ft (14 Events) 54
Figure 45. The Current and Proposed behind Start of Take-off Roll Directivity Adjustments... 56
Figure 46. The Differences between the Current and Proposed behind Start of Take-off Roll Directivity Adjustments for Jet and Turboprop Aircraft 57
Figure 47. L_{AE} Contours for the Boeing 737 Jet Aircraft with the Current and Proposed behind Start of Take-off Roll Directivity Adjustments for Jet (70 to 85 dB L_{AE} in 5 dB increments) 63
Figure 48. L_{AE} Contours for the Beech 1900D Turboprop Aircraft with the Current and Proposed behind Start of Take-off Roll Directivity Adjustments for Turboprops (70 to 85 dB L_{AE} in 5 dB increments) 63
Figure 49. L_{DN} Contours at Airport A2 with the Current and Proposed behind Start of Take-off Roll Directivity Adjustments (55 to 75 dB L_{DN} in 10 dB increments) 65
Figure 50. Volpe NoiseLogger™ System Setup.. 72
Figure 51. Airbus A310 Directivity Pattern at 1950 ft (1 Event) 73
Figure 52. Airbus A319 Directivity Pattern at 1950 ft (16 Events).. 73
Figure 53. Airbus A320 Directivity Pattern at 1950 ft (48 Events).. 74
Figure 54. Airbus A330 Directivity Pattern at 1950 ft (4 Events).. 74
Figure 55. Airbus A340 Directivity Pattern at 1950 ft (5 Events).. 75
Figure 56. Boeing 717 Directivity Pattern at 1950 ft (3 Events) 75
Figure 57. Boeing 737 Directivity Pattern at 1950 ft (20 Events) 76
Figure 58. Boeing 747 Directivity Pattern at 1950 ft (7 Events) 76
Figure 59. Boeing 757 Directivity Pattern at 1950 ft (12 Events) 77
Figure 60. Boeing 767 Directivity Pattern at 1950 ft (27 Events) 77
Figure 61. Boeing 777 Directivity Pattern at 1950 ft (12 Events) 78
Figure 62. McDonnell Douglas DC9 Directivity Pattern at 1950 ft (23 Events) 78
Figure 63. Bombardier CL600 Directivity Pattern at 1950 ft (104 Events) 79
Figure 64. Turbo-Propeller Aircraft Directivity Pattern at 1950 ft (14 Events)......................... 79

1 INTRODUCTION

The National Aeronautics and Space Administration (NASA), Langley Research Center (LaRC) and the Environmental Measurement and Modeling Division of the Department of Transportation's Volpe National Transportation Systems Center (Volpe) conducted a noise measurement study to examine aircraft sound level directivity patterns behind the start-of-takeoff roll. The study was conducted at Washington Dulles International Airport (IAD) from October 4 through 20, 2004.

1.1 Background

The Society of Automotive Engineers (SAE) Aircraft Noise Committee (A-21) initiated an activity to update the behind start of take-off roll algorithm in SAE AIR 1845, Procedure for the Calculation of Airplane Noise in the Vicinity of Airports[1], which is based on the 1980 report, Analysis of Selected Topics in the Methodology of the Integrated Noise Model[2]. AIR-1845 is the foundation on which the Federal Aviation Administration's (FAA) Integrated Noise Model (INM[3,4]) is based, as well as other aircraft noise prediction models. The INM, first released in 1978, is a software program designed to model noise in the vicinity of airports. It will be replaced by the FAA's next generation of environment modeling tool; the Aviation Environmental Design Tool (AEDT), which is currently under development. Both AEDT and INM model take-off noise using two parameters; (1) aircraft- and operation-specific noise data; and (2) a fleet-average behind start of take-off roll (SOTR) directivity adjustment. The aircraft and operation-specific noise data consist of noise-power-distance (NPD) data and one-third octave-band spectral data for each aircraft in the INM/AEDT database that are routinely updated by aircraft manufacturers and acoustic consultants, and have been strictly vetted through a verification and validation process. The SOTR directivity adjustment was not aircraft-specific, and it was based on now out-of-date aircraft fleet data.

The current SOTR directivity adjustment is represented in the following equations:

For $\quad 90° \leq \theta < 148.4°$

$$DIR_{ADJ} = 51.44 - (1.553\,\theta) + (0.015147\,\theta^2) - (0.000047173\,\theta^3) \quad (1)$$

For $\quad 148.4° \leq \theta < 180°$

$$DIR_{ADJ} = 339.18 - (2.5802\,\theta) - (0.0045545\,\theta^2) + (0.000044193\,\theta^3) \quad (2)$$

The SOTR directivity adjustment is 0.0 dB for azimuth angles between 0 and 90 degrees, and is symmetrical on either side of the aircraft. The current SOTR directivity adjustment is shown in Figure 1. The graphic presents aircraft sound level data corrections to be applied during SOTR as a function of polar angle and distance behind the aircraft. The pattern shows a "notch" of decreased sound level directly behind the aircraft which smoothes out as a function of distance behind SOTR. Note also that the directivity pattern is symmetrical about the axis defined by the aircraft fuselage.

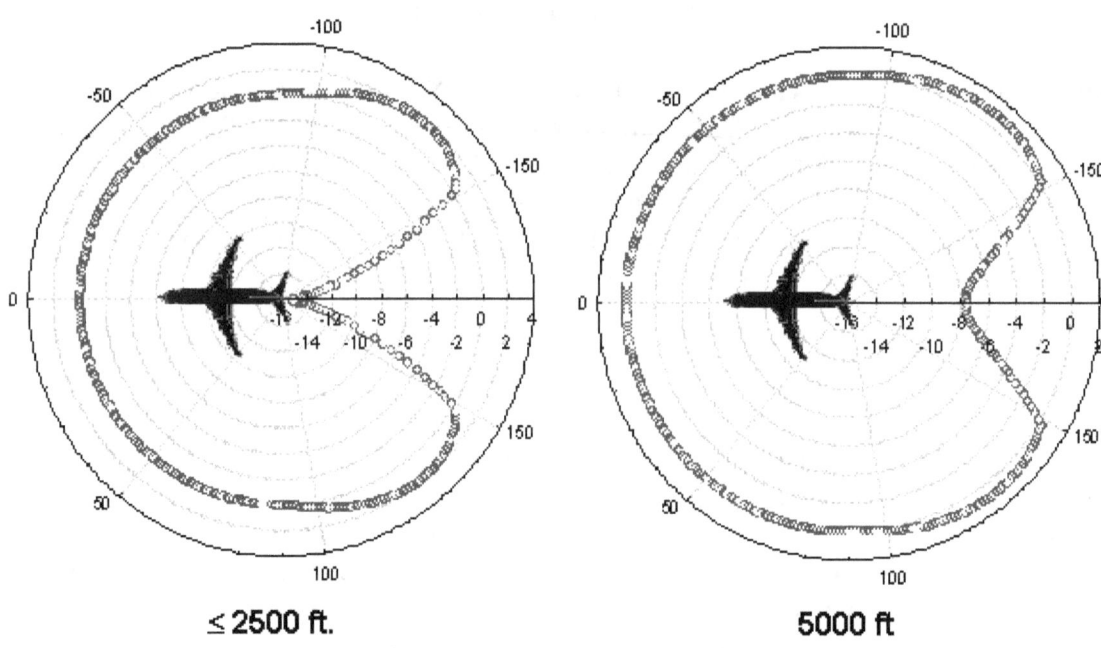

Figure 1. Plots of Current Directivity Algorithm

In INM/AEDT the SOTR directivity adjustment is applied to the noise levels associated with the first segment of an aircraft departure according to the azimuth angle behind the aircraft. It is also applied to noise from aircraft run-up operations. In addition, a smoothing algorithm is applied to the noise levels for receivers at distances behind the start of take-off roll greater than 2500 ft.

The current algorithm is based on measurements of older-generation, low by-pass ratio aircraft, primarily configured with fuselage-mounted engines. Since most modern jet aircraft are configured with wing-mounted engines, which tend to have higher by-pass ratios as well as operate at higher thrust levels, measurements to collect updated data were needed to better represent modeling of the current and anticipated future aircraft fleet. Additionally, directivity data for 4-engine commercial jet aircraft (which was limited to a single 747 event in the 1980 study), as well as propeller-driven aircraft were needed.

In October 2004, the National Aeronautics and Space Administration (NASA), Langley Research Center (LaRC), with support provided by the Volpe National Transportation Systems Center (Volpe), conducted a study at Dulles International Airport (IAD) to update the directivity algorithm in AIR-1845. In conjunction with the joint NASA/Volpe Behind Start of Take-off Roll study, the FAA's Partnership in Aviation Noise and Emissions Reduction (PARTNER) Center of Excellence (COE) conducted a simultaneous low frequency noise study at IAD. The low frequency noise study was designed to determine the effects of aircraft take-off noise on nearby structures as well as to study thrust reverser noise[5]. Although the NASA/Volpe and COE study shared instrumentation and ultimately utilized some of the same data, this report focuses on the NASA/Volpe study.

1.2 Objective

The objective of the joint NASA/Volpe study was to modernize the behind start of take-off roll sound level directivity algorithm in AIR-1845. This was accomplished by conducting field measurements at one or more large U.S. airports with a large variety of aircraft, analyzing the data to confirm and/or update the AIR-1845 algorithm, and recommending updated algorithms, potentially dependent on several variables, including aircraft-type.

2 SITE SELECTION AND LOGISTICS

2.1 Candidate Airport Selection

A total of 983 U.S. commercial airports were evaluated as potential measurement sites. In selecting candidate sites, several key factors were considered, including airport operations, fleet mix, runway configuration, and practical field measurement considerations, e.g., geographic location, accessibility and site terrain. Note that each factor has an associated set of benefits and/or detriments. For example, to minimize contamination from other aircraft, airports with a single runway might be considered ideal. The tradeoff in this example case would be the likely decrease in overall airport operations (and possibly fleet mix) due to the existence of only a single runway. To select appropriate candidate airports for measurements, key factors and tradeoffs were evaluated. This section presents a list of prioritized, candidate measurement sites and the methodology used to identify these sites. Based on the considerations documented herein, the goal was to identify at least ten candidate airports.

In selecting potential measurement sites, characteristics of each airport were examined and evaluated. These characteristics, which were used to narrow the list of potential measurement sites, include:

1. Geographic Location
2. Airport Elevation
3. Airport Fleet Mix
4. Site Terrain
5. Runway Configuration
6. Meteorological Conditions
7. Other Factors (i.e., NASA acoustic van accessibility, runway usage, etc.)

2.1.1 Geographic Location

Airports in Alaska and Hawaii were removed from the list of potential measurement sites to minimize obvious travel and logistical costs.

2.1.2 Airport Elevation

Airports at elevations of greater than 2,000 ft were removed from the list of potential sites to minimize the effect of altitude on aircraft performance.

2.1.3 Airport Fleet Mix

Airport fleet mix data were obtained by querying departure operations for 2003 from the Official Airline Guide (OAG) schedule. Fleet mix was examined for each airport. Three main types of aircraft were considered in rating airport fleet mix, starting with the highest priority: 1) Large commercial jet aircraft with wing-mounted engines such as the Boeing 737 and Airbus A320; 2) Large commercial jet aircraft with fuselage-mounted engines such as the McDonnell Douglas DC9; and 3) Four engine commercial jet aircraft and miscellaneous aircraft, such as the Boeing 747, Airbus A340, Embraer and Bombardier Canadair Regional Jet. Consideration was also given to airports with a wide variety of turboprop operations. Fleet mix for each airport was then evaluated and rated using the following method:

Table 1. Fleet Mix Rating Methodology

Priority	Aircraft Category	Rating	Points
High	Large jet aircraft with wing-mounted engines	Excellent	9
		Average	6
		Poor	3
Medium	Large jet aircraft with fuselage-mounted engines	Excellent	6
		Average	4
		Poor	2
Low	Four engine jet aircraft and miscellaneous aircraft, including turboprops	Excellent	3
		Average	2
		Poor	1

Points for each category were then totaled to obtain an overall fleet mix score. Airports that were considered to have high fleet mix scores were selected for final screening.

After considering fleet mix and removing airports due to elevation and geographic criteria, candidate airports were reduced to 48.

2.1.4 Site Terrain, Near Runway

Figure 2 presents ideal microphone locations, relative to a conceptual runway end, for behind start of take-off roll measurements. In general, the most complete microphone array would be located at the 1,000 ft radial. Fifteen microphones, each separated by 15 degrees, would be positioned along the 1,000 ft radial as shown in the illustration.

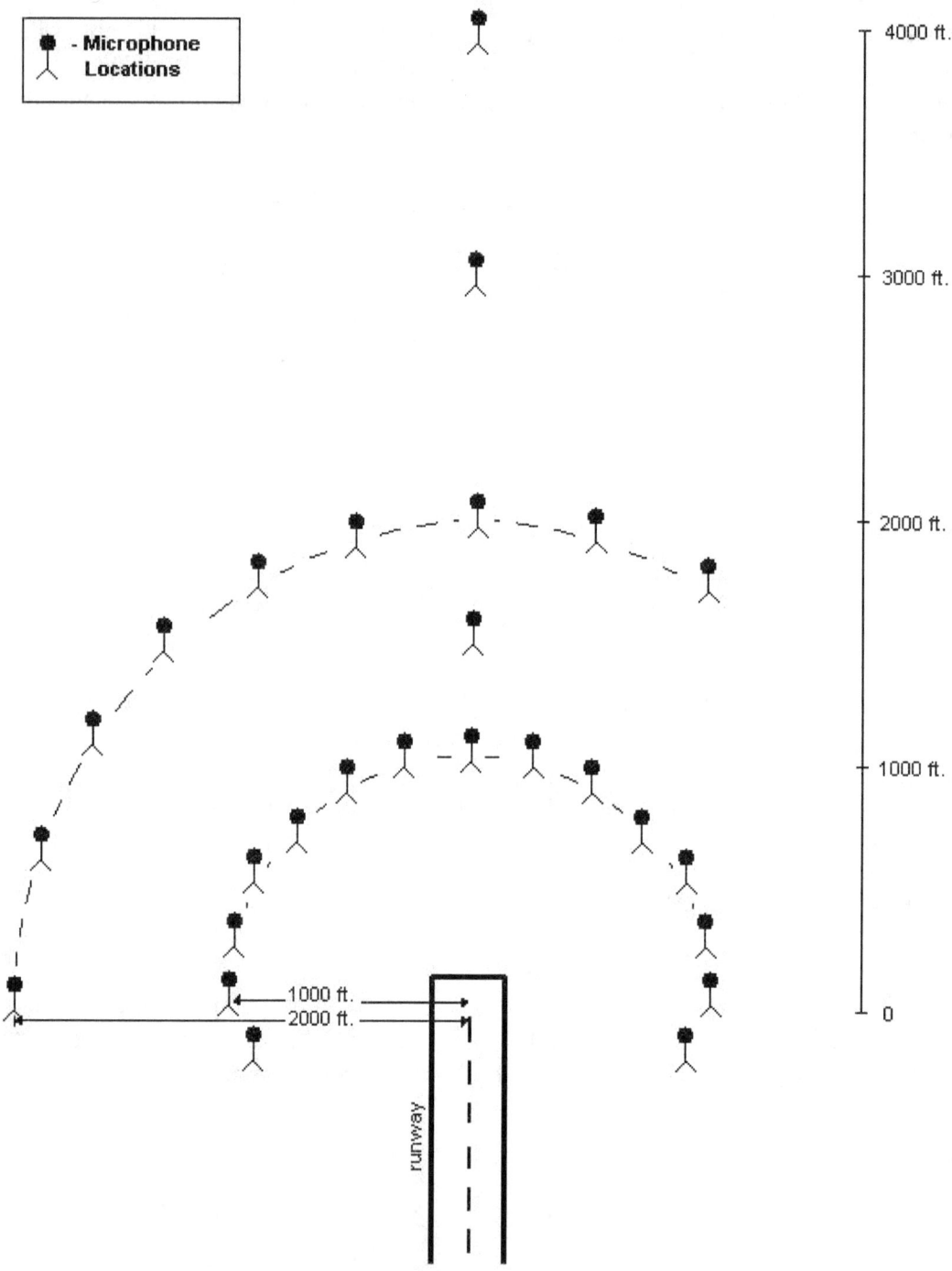

Figure 2. Ideal Microphone Array

Candidate measurement sites that allowed for the placement of microphones along a radial of approximately 1,000 ft behind the runway were considered acceptable. Additionally, an unimpeded line of sight from the microphone arc to the start of the runway was desired.

Using USGS aerial photos of the 48 candidate airports and a modified version of standard Geographic Information Systems (GIS) software, a semi-circle of 1,000 ft radius was superimposed on the start of each candidate runway. Figure 3 presents a representative photo for Runway 35L at Austin-Bergstrom Airport.

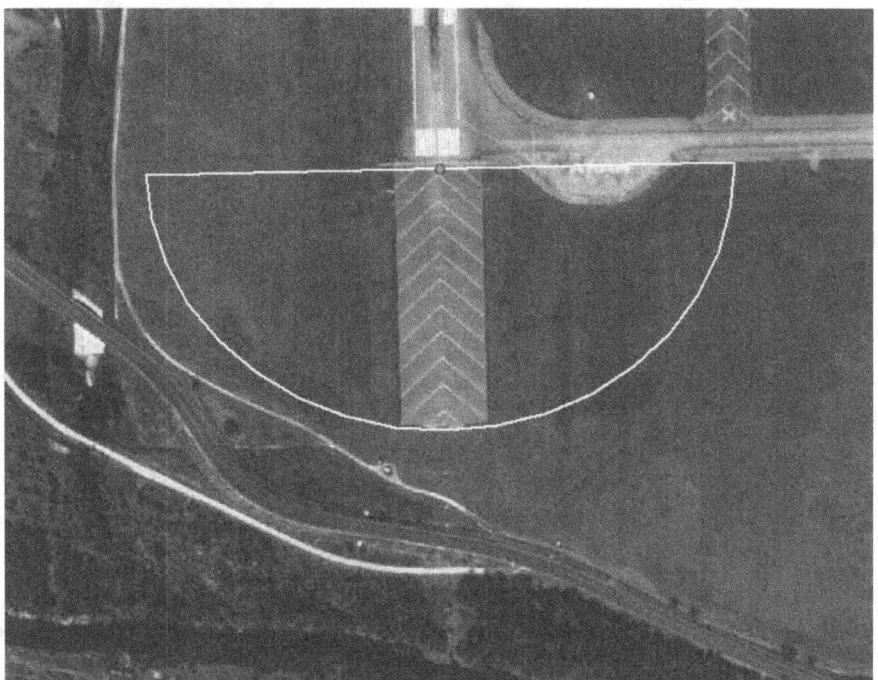

Figure 3. Aerial Photo of Runway 35L at Austin-Bergstrom Airport

The semi-circle behind the runway represents the nominal microphone line. The area within the semi-circle was visually surveyed for obstructions such as buildings, parking lots, heavy vegetation, bodies of water, and noticeable changes in elevation. This was done for runway ends at each of the remaining 48 candidate airports. This process also revealed other sources of practical problems, including highways, roadways, cross-runways and taxiways in close proximity.

Figure 4 shows an example of what was considered to be an ideal measurement site. The area behind the runway is obstacle free within a 1,000 ft radius and does not appear to have other noise sources in the immediate vicinity.

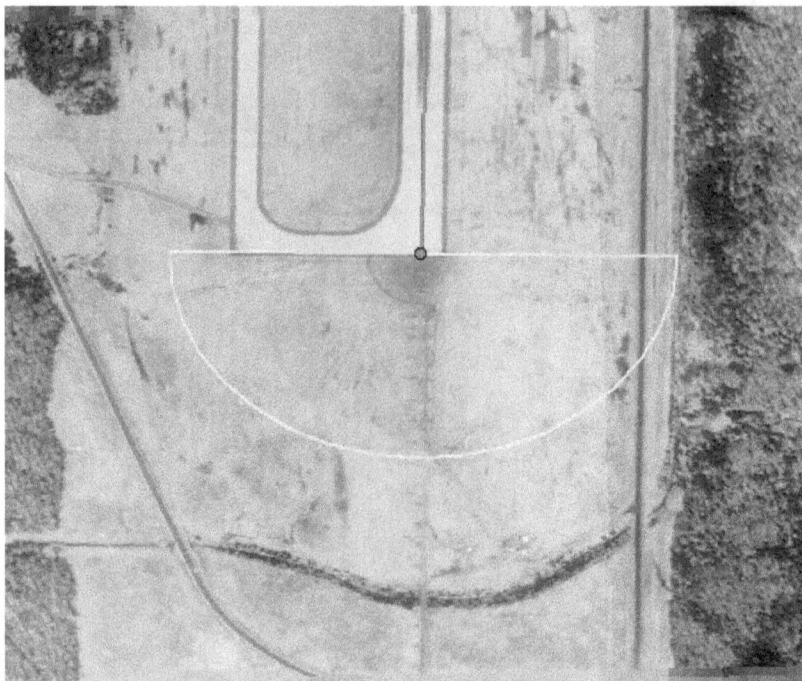

Figure 4. An Aerial Photo of Runway 1R at Dulles International Airport

Figure 5 shows a runway that was eliminated as a potential measurement site, given the close proximity of a major roadway, building and what appears to be a parking lot.

Figure 5. An Aerial Photo of Runway 9 at Tampa International Airport

2.1.5 Site Terrain, Distant

The need to quantify the directivity characteristics, as a function of increasing source-to-receiver distance, necessitated the addition of microphones at locations on semicircles 2,000 to 4,000 ft behind the runway. As depicted in Figure 2, nine microphones, also separated by 15 degrees, would ideally be utilized at 2,000 ft in order to help quantify aircraft directivity characteristics as a function of distance. "Centerline" microphones would also be placed directly behind the aircraft at approximately 1,500, 3,000 and 4,000 ft to further quantify the distance dependence. Accordingly, red semi-circles with radii equal to the 2,000 and 4,000 ft distances were superimposed on runway-end aerial photographs as shown in the example in Figure 6.

Figure 6. Aerial Photo of Runway 17R at Austin-Bergstrom Airport

As with the 1,000 ft semi-circle (in yellow), areas within the red semi-circles were visually inspected for obstructions. However, the entire area within the red semi-circles did not necessarily have to be obstruction free. As seen in Figure 2, microphones were only anticipated for placement at limited positions along the 2,000 ft semi-circle and at centerline locations at 1,500 ft, 3,000 ft and 4,000 ft. Only areas where microphones were planned for deployment were evaluated. Depending on the results obtained at 1000 ft, it was hypothesized that the data measured on one half of the 2,000 ft semicircle could be "reflected" over the centerline axis, if necessary i.e., the directivity pattern is symmetrical to the left and right side of the aircraft.

After completing an aerial photo analysis of the local terrain at near and distant runway ends, 18 additional airports were eliminated, leaving 30 potential candidate sites.

2.1.6 Runway Configuration

Any candidate airport runway that had another runway within a 5,000 ft arc behind it was eliminated because of: 1) potential noise contamination from aircraft on the cross runway, 2) the inability to place microphones on the cross runway, and 3) the inability to run cables across the cross runway. Airports were then grouped by number of eligible runways, where fleet mix scores were re-evaluated in the final selection of candidate airports. Although particular runways were eliminated, none of the 30 candidate airports were removed from the list.

2.1.7 Other Factors Considered

Due to their ability to collect 10 channels of time-synchronized acoustic data, the NASA acoustic vans were considered strong candidates for use in the measurement study. The vans, however, require a measurement site that is free of roadways, runways, and taxiways, to facilitate cable runs between the vans and microphones. Microphone systems at locations the NASA vans cannot access were to be instrumented with Volpe portable acoustic measurement systems.

It was also necessary to evaluate airport operations data to determine runway usage specific to departures. The OAG data do not provide runway usage information, simply total airport operations. It was possible for a candidate runway to be used mainly for aircraft arrivals, which would make the runway less desirable for this study. Therefore individual airport tower logs were needed.

2.1.8 Site Selection Results

Airports were ranked by fleet mix and number of eligible runways. The top ten candidate airports for measurements were as follows:

1. Seattle-Tacoma International (SEA), Washington
2. Cincinnati/Northern Kentucky (CVG), Kentucky
3. Dulles International (IAD), Washington D.C.
4. Portland International (PDX), Oregon
5. Orlando International (MCO), Florida
6. Minneapolis-St. Paul International (MSP), Minnesota
7. Raleigh-Durham International (RDU), North Carolina
8. Kansas City International (MCI), Missouri
9. Indianapolis International (IND), Indiana
10. Charlotte/Douglas International (CLT), North Carolina

2.1.9 Meteorological Data

In general, outdoor acoustic measurements require minimal winds in the vicinity of the measurement site. Accordingly, meteorological data were obtained and evaluated for the top ten candidate airports. Surface meteorological data for the years 1984, 1989, and

1992 were downloaded from an EPA website[6]. The data provided hourly averaged wind speeds throughout the year. The wind data for the three years were then averaged. Results showed that winds at the ten airports were relatively similar. None of the candidate airports were eliminated as a result of the wind speed data.

2.1.10 Selection of Study Airport

In the above list of candidate airports, IAD was considered the preferred choice, given Volpe's familiarity with the airport due to recent measurements, the team's knowledge of key personnel at IAD, and Langley's relatively close proximity to the airport.

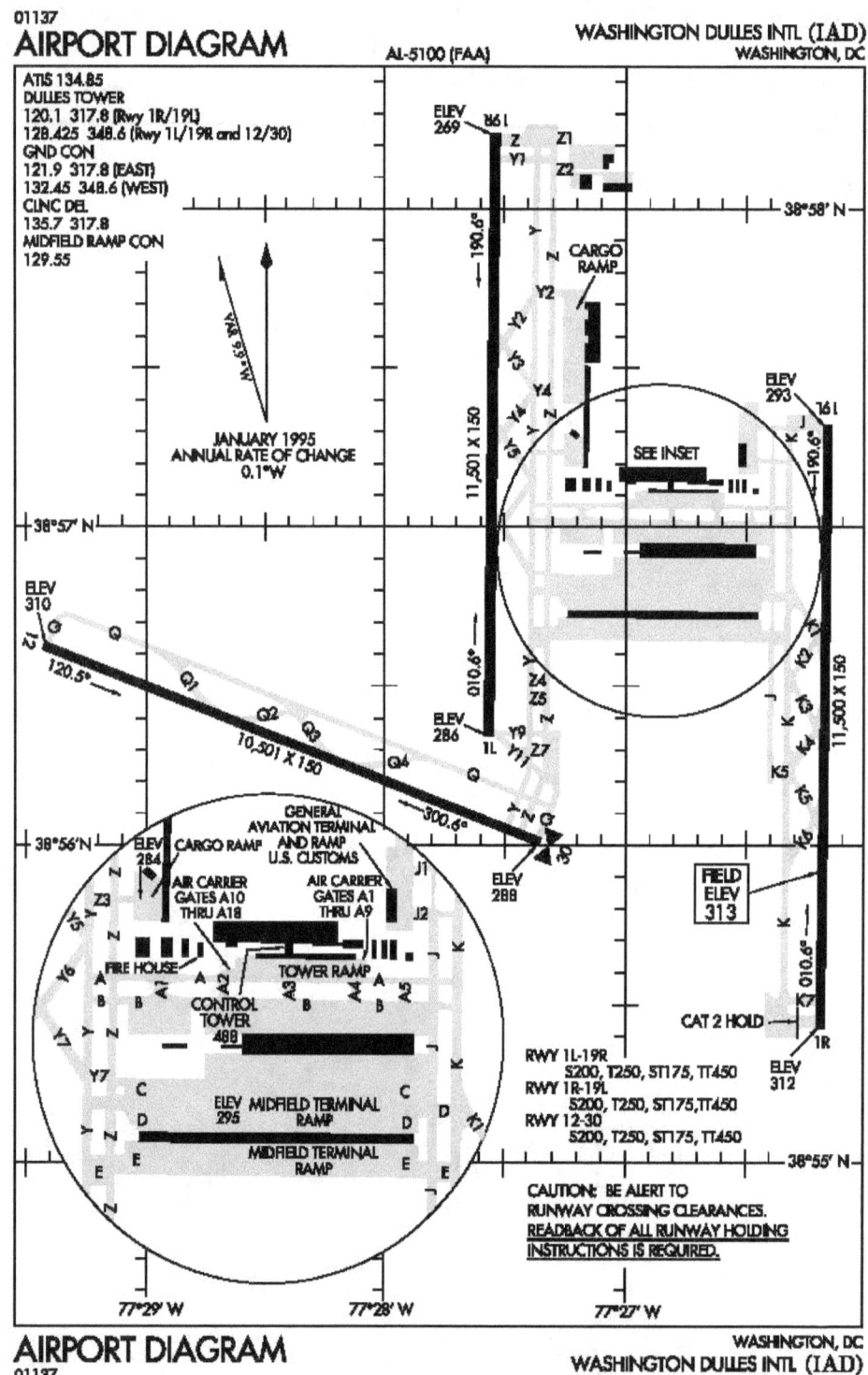

Figure 7. Airport Diagram of Dulles International Airport (IAD)

2.2 Measurement Site Logistics

2.2.1 Primary Runway for Measurements

Subsequent to approval from airport management to conduct the study, further communication with appropriate personnel, specifically IAD Air Traffic Control (ATC) was conducted. The purpose of the communications was to gather information regarding runway usage as discussed in Section 2.1.7. Information on airport traffic operations at IAD (Table 2) led to the conclusion that Runway 30 would be the target runway for measurements. These data suggest that the majority of aircraft take-off operations occur on Runway 30.

Table 2. IAD Runway Usage Data

Condition	How often (rarely, sometimes, mostly) ?	1L	1R	19L	19R	12	30
Landing to he North (high departures demand)	mostly	L	L/T	x	x	x	T
Landing to he North (high arrivals demand)	mostly	L	L				T
Landing to he South (high arrivals demand)	mostly			L/T	L	L	
Landing to he South (high departures demand)	mostly			T	L	x	T
Landing to he South (winds prohibit 30 departs.)	sometimes			L/T	L/T	x	x
Landing to he North (strong NW winds)	rarely	x	L				L/T
Landing to he North (strong NW winds)	rarely	x	x				L/T
Landing to the South (strong E-SE winds)	extremely rare			x	x	L/T	

T = Take-off
L = Landing
x = not used

2.2.2 Aircraft Departure Schedule

IAD airport operations data for a week in October were obtained from the 2003 OAG schedule. The data, presented in Figure 8 suggest that peak operations occur during the time blocks of 8:00-10:00, 12:00-14:00, 16:00-18:00, and 20:00-22:00. Operations data obtained from Federal Aviation Administration's System for assessing Aviation's Global Emissions (SAGE)[7] showed generally consistent trends in departure operations. Observations during the preliminary site survey (described in the next Section) also confirmed the overall trend.

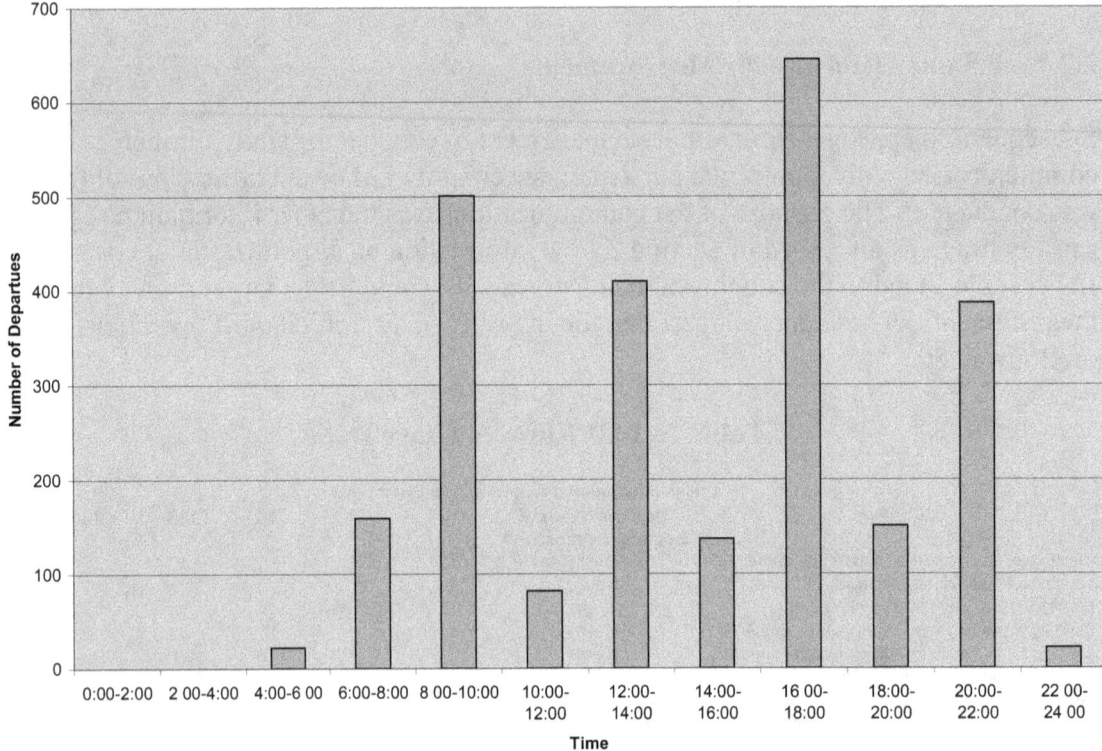

Figure 8. Departure Flights at IAD During 10/1/03-10/7-03

Wing-mounted 4 engine aircraft were of particular interest in the study. Figure 9 indicates that IAD generally has a small number of Boeing 747 departures. Figure 10 also shows a small number of flights for the Airbus A340. The majority of the wing-mounted, 4 engine aircraft departures in October (i.e., the planned month for the study) occurred during the time period of 16:00 to 22:00.

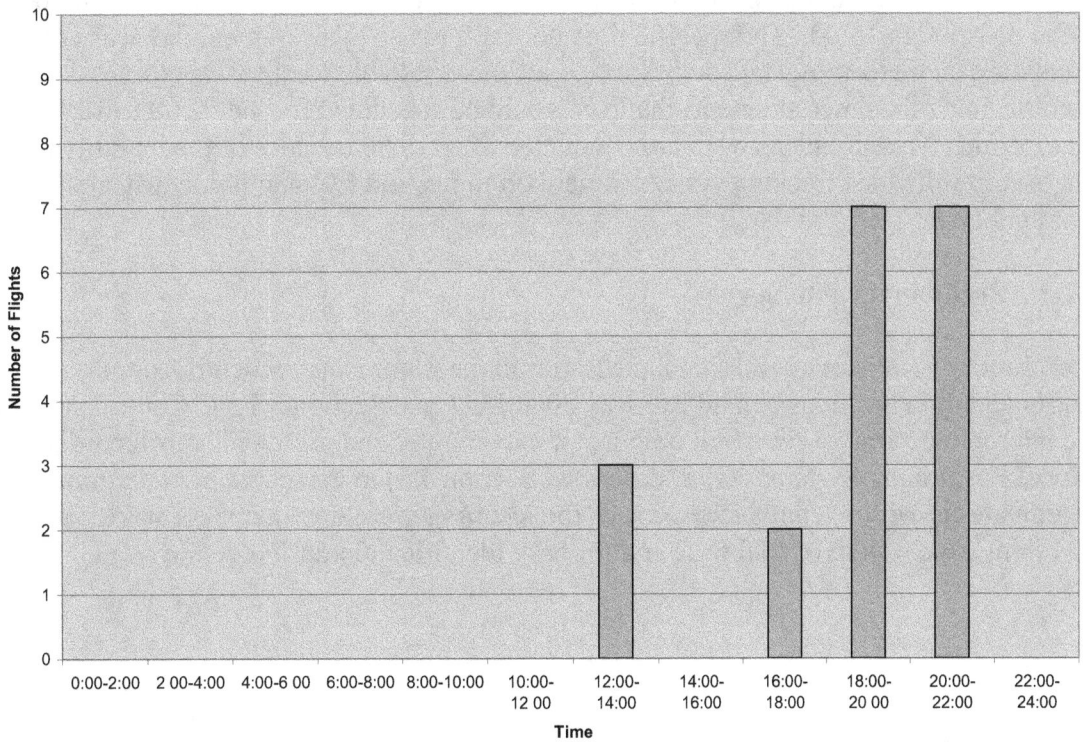

Figure 9. Boeing 747 Departure Flights at IAD During 10/1/03-10/7-03

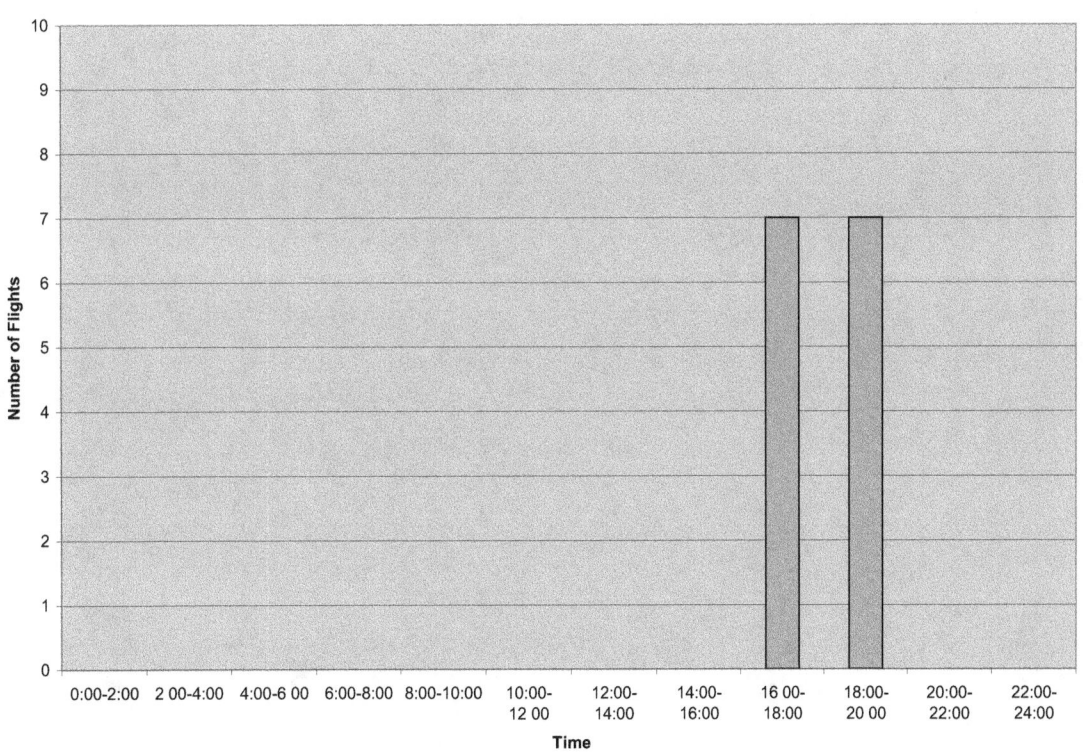

Figure 10. Airbus A340 Departure Flights at IAD During 10/1/03-10/7-03

As a result of these findings, measurements were targeted for the time periods of 8:00 to 14:00 or 12:00 to 18:00. The specific time period for each measurement day was chosen depending on meteorological conditions e.g., if more suitable weather was forecast in the evening hours then measurements that day would be scheduled for the 12:00 to 18:00 time period. Measurements were not conducted after 18:00 because of the inability to use the aircraft video tracking system (described in Section 3.2) and to identify aircraft tail numbers.

2.2.3 Preliminary Site Survey

A preliminary site survey was conducted prior to measurements to identify site instrumentation locations behind Runway 30 at IAD (see Section 3.1.2). Coordinates of the microphone array were saved onto a laptop computer and used with a differential Global Positioning System (dGPS, described Section 3.5) to locate the predetermined coordinates to within a few inches. Once the locations were found, a short stake with surveying tape was fixed to the ground for easy identification during deployment.

3 INSTRUMENTATION

The primary instrumentation used during measurements is listed in Table 3. For clarity, instrumentation specifications and locations are described in more detail in Sections 3.1 through 3.5. Measurement procedures are described in Section 4. A detailed layout of the instrumentation used in the study can be found in Appendix A. Angle, X and Y coordinates of the instrumentation locations are based on the local Cartesian coordinate system as described in Section 3.1.2

Table 3. Summary of Primary Instrumentation

Instrumentation	Quantity	Responsible Organization	Purpose	Comments
DGPS System	1	Volpe	Survey of instrumentation locations	In addition to maps and aerial photos, the dGPS system was used to survey the location of instrumentation to an accuracy of better than a meter
Acoustic Vans	2	NASA	Collection of acoustic data	Microphone system (up to 10 microphones per van)
Portable Acoustic System	8	Volpe	Collection of acoustic data	Portable microphone system
Digital Video Tracking System	2	Volpe	Tracking of aircraft position and acceleration profile	This system will define the break release point, liftoff point and acceleration profile
Portable Weather Station: TAMS System	1	Volpe	Collection of position specific meteorological data	This information will supplement the airport tower meteorological data and allow for a much more detailed definition of meteorological

3.1 Acoustic Instrumentation

Acoustic instrumentation used during measurements consisted of a combination of NASA and Volpe microphone systems.

The NASA microphone system consisted of two separate vans for acoustic data acquisition. NASA Van 1, designated as GMC/19, was deployed to collect data from nine microphones in the 900 ft array and the centerline microphone at 1350 ft. NASA Van 2, designated as Barth/72 (Figure 11), was deployed to collect data from six microphones in the 1950 ft array, two microphones at 2500 ft, and one microphone at 3450 ft Portable Volpe acoustic measurement systems were deployed to facilitate data collection at remote locations.

3.1.1 Specifications

3.1.1.1 NASA Acoustic Vans

Each of the NASA acoustic vans were outfitted with a Digital Acquisition Measurement System (DAMS) capable of collecting ten independent channels of acoustic data. For each channel of the system, the signal was digitized at the microphone, transmitted via cable to the van, multiplexed with time and test run information, and then recorded on a hard disk drive. The sample rate at each microphone was 25 kHz with the anti-aliasing frequency set at 12.5 kHz. The data were transferred from the hard disk drive to a PC for signal processing. The digital acoustic time domain data were transformed to the frequency domain at one second intervals using the average of eleven 4096-point fast Fourier transforms (FFTs) with a Hamming window and 50 percent overlap applied, resulting in 0.983-second blocks of data. These FFTs were used to compute narrowband data, which were converted to 1/3-octave band data. The processed data were stored on digital video disks (DVD) for later off-line reduction and analysis.

The NASA microphones were the Bruel and Kjaer (B&K) Model 4134. These are condenser microphones requiring a polarization voltage. The polarization voltage was supplied by either a B&K Model 2669 or 2619 preamplifier and a modified version of the B&K Model 2804 power supply. Battery powered booster boxes, consisting of a transceiver in and a transceiver out, provided polarization voltage to microphones at far distances. Digitization of the acoustic signal was performed using a Burr-Brown Model ADC 76KG A-to-D converter. A B&K Model 0238 3.5 inch (9 cm) diameter foam windscreen was placed atop each microphone to reduce the effects of wind-generated noise on the microphone diaphragm. Microphones were installed on tripods at a height of 4 ft above ground level (Figure 12).

Figure 11. NASA Acoustic Van 2, Designated as "Barth/72"

Figure 12. NASA Microphone System with Power Booster Box

3.1.1.2 Volpe Acoustic Measurement Systems

The Volpe acoustic measurement system is a specialized measurement system, which may be used to conduct unattended, long-term (30+ days) and continuous 1/3-octave band noise measurements in outdoor environments. The system is designed to be compact, light, rugged, and can run for 96 hours with external battery power, or continuously with solar panels, thus, making the system able to store significant amounts of data. The system uses G.R.A.S. Model 40 AE or Model 40AQ ½-inch electret microphones and is powered by a G.R.A.S. Model PRM902 preamplifier. A B&K Model 0237 3.5 inch (9 cm) diameter foam windscreen was placed atop each microphone. Each microphone is interfaced to a Larson Davis Model 824 real-time analyzer/sound level meter. Data are stored on an Itronix Husky Fex21 handheld computer. The system also includes a FT Technologies Model 702 ultrasonic anemometer to measure wind speed and direction. The output of the anemometer is also stored on the handheld computer. The system collected 1/3-octave acoustic data along with wind speed and direction data at 1-second intervals. Both the microphones and anemometers were installed on tripods at a height of 4 ft above the local ground surface (Figure 13). Appendix D displays in more detail a diagram of the Volpe acoustic measurement system set up.

Figure 13. Volpe Acoustic Measurement System

3.1.2 Locations

Runway 30 has three taxiway turning points (Y, Z, Q in Figure 14). The turning point used for the majority of take-offs is crucial since it would be ideal to define the origin of the microphone array based on the location of the start of take-off roll for the majority of operations. IAD air traffic control indicated that taxiway turning point Q is used most frequently during departures.

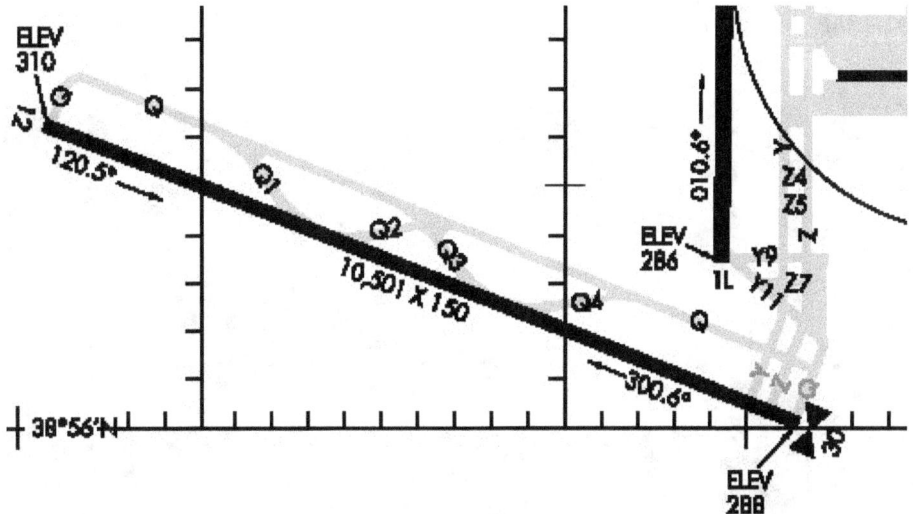

Figure 14. Taxiway Turning Points Y, Z, Q on IAD Runway 30

As a result, a local Cartesian coordinate system (x, y, z) was established with the origin (0, 0, 0) at 300 ft from the end of Runway 30, where aircraft taxiing from turning point Q were observed to begin take-off roll. The centerline of Runway 30 was designated as the x-axis such that positive x values increase throughout the ground roll.

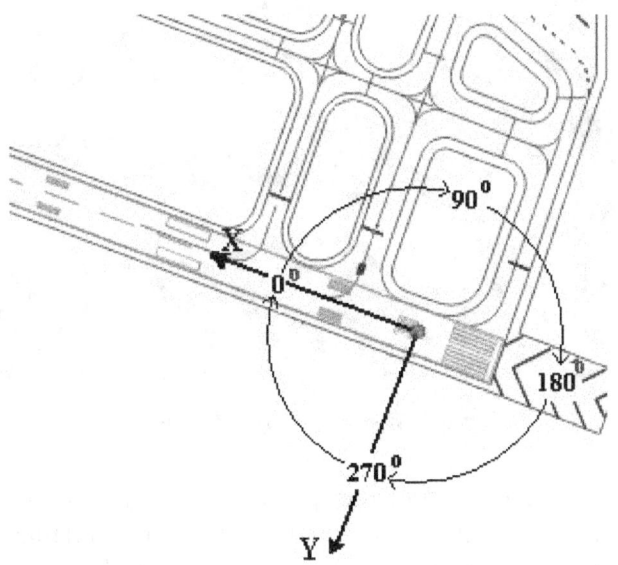

Figure 15. Local Cartesian Coordinate System Used on Runway 30

Local coordinates were calculated for the corresponding microphone locations depicted in Figure 2. Coordinates for alternate microphone locations were also calculated in case a site survey led to some revisions of microphone locations e.g., a new location was necessary if the preferred location was obstructed from view of the runway. As a result of this precaution, "zones" of where microphones could be placed were superimposed onto an aerial photograph of Runway 30. Figure 16 shows a range of potential microphone locations. The semicircle zones are 200 ft wide and are centered at 1,000, 2,000, and 4,000 ft radii from the origin.

Figure 16. Microphone Location Zones Behind Runway 30

The primary and a set of alternate coordinates were then saved onto a laptop computer and used in conjunction with dGPS to identify and mark instrumentation positions (see Section 2.2.3). The microphone location zones proved useful when a site survey discovered unforeseen terrain complications and obstructions at 1000 and 2000 ft. The 1000 and 2000 ft radials were moved to 900 and 1950 ft, respectively. A taxiway at the left side of the arc on Runway 30 disallowed the placement of microphones at 90 and 105 degrees. The microphone at 900 ft and 135 degrees, designated as N2, was placed in a location with an approximate 5 ft depression. Line-of-sight blockage was observed at this location. As a precaution, an additional microphone, N2A, was placed at 900 ft and 140 degrees, which did not have line-of-sight blockage.

As mentioned in Section 2.1.5, centerline microphones were to be ideally placed directly behind the aircraft at approximately 1,500, 3,000 and 4,000 ft to quantify the drop-off in sound level. However, to assist the FAA PARTNER COE low frequency noise study, the 3,000 ft centerline microphone was moved to a location at 3,600 ft. As a result, the 4,000 ft centerline microphone was moved to 4,600 ft. An additional centerline microphone was also added at 2,600 ft. The resulting IAD, Runway 30-specific microphone setup is shown in Figure 17.

Figure 17. Final Instrumentation Locations

3.2 Aircraft Tracking System

A digital video tracking system was used to determine location of the aircraft brake release point, point of lift-off and aircraft acceleration profile. The system utilized the local Cartesian coordinate system discussed in the previous Section. The X, Y, and Z position of the aircraft must be known to track the aircrafts' movement on the runway. The assumption that the aircraft followed the runway centerline during take-off roll indicated that the Y-coordinate would be at a constant value of zero. The digital video tracking system is capable of defining the X and Z coordinates of the aircraft during take-off roll.

Figure 18. Digital Video Tracking System

3.2.1 Specifications

The digital video tracking system uses two Canon Optura digital video cameras. The cameras have a 720 by 480 pixel Charged Couple Device (CCD) and create a video image stored on the mini-DV tape medium. Installed on the cameras are Kenko wide-angle lenses (Model VC-050Hi), which allow for a usable viewing angle of approximately 60 degrees.

For determination of an aircraft acceleration profile, the tracking system required a time coordination sub-system. The time coordination sub-system is comprised of a TrueTime time code generator (Model 705-326), a laptop computer, and a LED box mounted in view of the camera. The basic principle of the timing coordination system is that for every tracked object, the exact time of all the frames in both cameras must be known. To accomplish this, DOS software was developed and installed onto a laptop computer to communicate with the TrueTime time code generator and LED box. The DOS software instructs the time code generator to send a precisely known timing pulse to the LED box. When the time code generator receives the signal, a voltage is sent to the LED box, the LED illuminates, and the camera captures this image. The DOS software records the time at which the timing pulse is sent to the LED box.

Optical targets are manually placed in the field of view of the cameras. Each camera must have two targets in its field of view during the entire event of interest. One target, referred to as the Direct Target, determines the pitch and yaw of the camera relative to the coordinate system. The other target, referred to as the Angle Target, is used together with the Direct Target to determine the roll of the camera relative to the coordinate system. An important feature of the system is that the cameras can be rotated between events of interests - as long as two targets are in view of the camera, the system software can determine the roll, pitch and yaw of the camera. The Optical Targets are 2 ft square plywood boards with alternating 1 ft black and white squares on one side and a support and mounting structure on the other side. The Optical Targets are designed to mount on top of standard consumer-grade photography tripods. When mounted, the Optical Targets were two to three feet above the ground.

3.2.2 Locations

A digital video tracking system camera (TRACK1 in Figure 17) was placed at approximately 1100 ft from the southern side of Runway 30's centerline. TRACK1 captured the aircraft at its brake release point within its field of view. A second digital video tracking system camera (TRACK2) was placed approximately 2500 ft from the northern side of Runway 30. This setup allowed a combined field of view of about 5,500 ft of the runway allowing the cameras to capture the entire take-off roll profile and lift-off points for the majority of aircraft types.

3.3 Portable Weather Stations

3.3.1 Specifications

A Qualimetrics Transportable Automated Meteorological Station (TAMS) was used to measure wind speed, wind direction, relative humidity, and air temperature at one-second intervals. A complete TAMS system consists of a sensor unit and a control/display unit that displays real time meteorological data. The battery-powered stations are portable and well suited for remote sampling. Wind speed can be measured from a stall speed of 2 mph, to a maximum of 55 mph, with an accuracy of 1 mph or 5% of range (whichever is greater) and a resolution of 1 mph. Wind direction can be measured a full 360 degrees with a root mean standard error of 18 degrees and a resolution of 10 degrees. Temperatures can be measured from –9 to 110 degrees Fahrenheit with an accuracy and resolution of 1 degree. Relative humidity is accurate to within 3%, with a resolution of

1%. The unit was placed on a tripod at a height of 4 ft. Data from the control unit is automatically saved onto a Hewlett Packard Model 200LX palmtop computer.

Figure 19. Transportable Automated Meteorological Station (TAMS)

3.3.2 Locations

TAMS was positioned at the Test Director's location (TD/ TL in Figure 17) so that wind speeds could be monitored before and during an event. The Test Director was able to easily determine if an event was good/no good based on pre-determined wind speed criteria.

3.4 Tail Logging Instrumentation

3.4.1 Specifications

Standard binoculars were used to log tail numbers during measurements. Aircraft tail numbers, event start/stop times, event identification, aircraft and engine types were handwritten onto a log sheet (see sample in Appendix C).

Figure 20. Primary Tail Number Logger

3.4.2 Locations

Two field personnel observed and logged the tail number of each test aircraft. The primary logger was positioned alongside the Test Director. A secondary logger was positioned at the digital video tracking system location south of Runway 30.

3.5 Other Instrumentation

The differential global positioning system (dGPS) was used prior to measurements to identify instrumentation locations to within a few inches. Section 2.2.3 and 3.1.2 discusses the use of this system.

Figure 21. Differential Global Positioning System (dGPS) Receiver Antenna

3.5.1 Specifications

The dGPS is designed around two single-frequency Novatel Model RSAAB SF3400E GPS receivers and two GLB Model SNTR150 transceivers. The two 25-watt GLB radio transceivers are tuned to a frequency of 136.325 MHz and together with a graphical user interface (GUI) installed on a laptop computer, allow system initialization, real-time position display, and data storage. More detailed specifications of the dGPS can be found in the Differential Global Positioning System User's Guide[8].

3.5.2 Locations

The dGPS system was used for conducting the pre-measurement site survey as discussed in Section 3.1.2.

4 MEASUREMENT PROCEDURES

The following section describes the procedures for a typical day of measurements. Measurements took place during October 2004 and consisted of 10 days of data collection. Figure 22 below summarizes the measurement schedule.

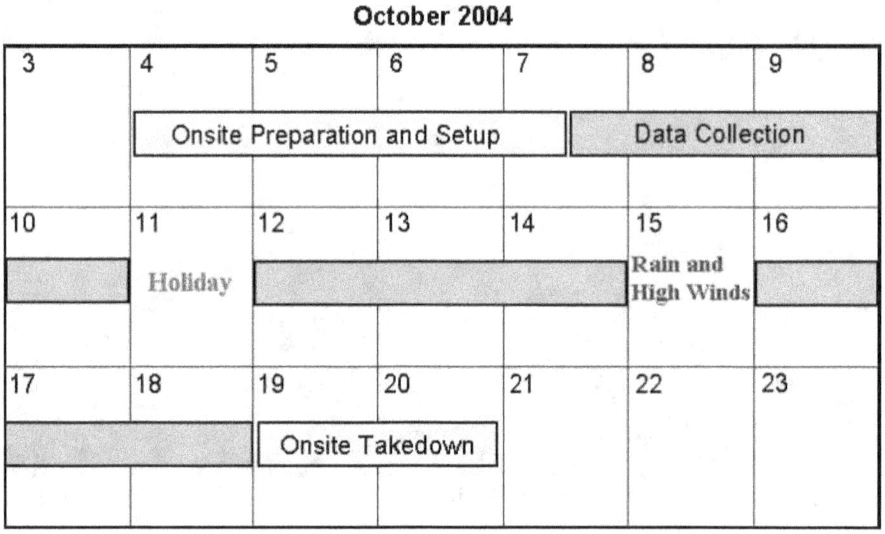

October 2004

3	4	5	6	7	8	9
	Onsite Preparation and Setup				Data Collection	
10	11	12	13	14	15	16
	Holiday				Rain and High Winds	
17	18	19	20	21	22	23
		Onsite Takedown				

Figure 22. Measurement Schedule

4.1 NASA Acoustic Vans

4.1.1 Deployment

Prior to measurements each day, a calibration tone comprised of a sine wave of 124 dB at 250 Hz was applied to each NASA microphone and the resultant signal was recorded. White noise was then passed through the system and also recorded. The system clocks were automatically time synchronized using GPS. The Test Director was notified once calibration was complete and that the acoustic vans were standing by for data collection.

4.1.2 During Measurements

When the Test Director announced the event number and "data on" (see Section 4.6.2) over 2-way radio, data collection commenced on both NASA acoustic vans. Field personnel from the vans recorded the event number. Real time data were monitored to make certain systems functioned normally and to detect potential contamination. At the end of each measurement day, the same calibration tone used at the beginning of the day was applied to the microphones and recorded to ensure the system's input sensitivity had not changed.

4.2 Volpe Acoustic Measurement Systems

4.2.1 Deployment

Prior to measurements each day, a calibration tone comprised of a sine wave of 94 dB at 1 kHz was applied to each portable acoustic system and the resultant signal was recorded. A microphone simulator, used to measure the noise floor of the system, was then applied to the system and recorded. The system clocks were then synchronized. Time synchronization was accomplished using a digital wristwatch, synchronized to the TrueTime time code generator used in the digital video tracking system (Section 3.2.1). Data acquisition began upon completion of calibration and time synchronization. The Test Director was then notified that calibration was complete and measurements with the portable acoustic systems had been initiated.

4.2.2 During Measurements

The portable acoustic measurement system recorded data continuously from deployment until the end of the measurement day. During periods where no measurements were made (e.g., when aircraft did not depart from Runway 30), functionality of the systems was periodically checked. These quality assurance checks consisted of a power check, cable connection check, and data recording check. At the end of the measurement day, the same calibration tone used at the beginning of the day was applied to each portable acoustic measurement system and recorded to ensure the system's input sensitivity had not changed.

4.3 Digital Video Tracking System

4.3.1 Deployment

Prior to measurements, a manual survey of the vertical positioning of the cameras and optical targets was completed. Using a transit (and its associated tripod and leveling devices) and a measuring stick, the vertical positions of the cameras and targets were measured. Measured positions were noted in a log sheet to be used for latter data processing.

At the beginning of each measurement day, camera systems were set up, heights of the center of the camera lens and targets were measured, and the clock on each camera was checked to ensure the correct time was displayed. The system laptop computer and the TrueTime time code generator were then set-up and the LED indicator was tested for functionality. Upon completion of deployment and testing, the Test Director was notified that deployment was complete and the Digital Tracking System was ready for data collection.

4.3.2 During an Event

When the Test Director transmits over two-way radio that an event was about to begin, camera operators immediately used the DOS program (see Section 3.2.1) to send an LED signal to the cameras. Operators confirmed the LED indicators and entered the event number into the laptop computer. The laptop computer automatically logged the time associated with the LED indicator.

4.4 Meteorological Station

4.4.1 Deployment

The TAMS meteorological system was set up at the Test Director's location. The system clock was synchronized using a TrueTime time code generator. The TAMS unit began collecting data immediately after the system was deployed.

4.4.2 During an Event

The TAMS system recorded data continuously from deployment until the end of the measurement day. Real time wind speed was also displayed on the control unit. The Test Director used the TAMS system as a guide to help judge whether an event was good/no good due to wind speeds. As recommended by AIR 1845, any event in which the wind speed exceeded 15 knots (approximately 18 mph) was considered no good.

4.5 Aircraft Tail Number Loggers

4.5.1 Deployment

Prior to measurements both tail number loggers positioned themselves in a location that provided clear view of the aircraft tail numbers. Once both loggers were in position, the Test Director was notified.

4.5.2 During an Event

During an event, both tail loggers noted the event start time, which was when the Test Director announced, "data on." The tail loggers also recorded the end time, event ID, tail number, airline, number of engines, body type, engine configuration, starting point (see Section 3.1.2), and type of take-off roll (see paragraph below) onto log sheets. The tail number loggers noted their best guess of the aircraft type as well. An example of the log sheet is included in Appendix C.

There were three types of start of take-off roll: rolling, static, and continuous as follows:

- A rolling start occurred when the aircraft reached the runway end to initiate take-off and came to a full stop. Brakes were then released and engine run-up began. The aircraft begins its ground roll as engine power was spooled up. In this scenario the aircraft may not achieve take-off thrust until several hundred feet down the runway as a result of the engines' spool-up time. This was the most common type of take-off roll.

- A static start occurred when the pilot applied full take-off power, holding the aircraft brakes until take-off power was achieved, then released the brakes and initiated ground roll. In this scenario, take-off power was achieved closer to the end of the runway. This was more common for larger aircraft such as the Boeing 747 and 777.
- The third type of take-off roll, continuous, occurred when the aircraft approached the runway but did not come to a full stop. Engine run-up was noticed midway during the aircraft's turn onto the runway. The aircraft remained in motion from the turning point and throughout the entire ground roll. In this scenario the aircraft may not achieve take-off thrust until several hundred feet after alignment on the runway, due to the combination of the aircraft moving and the engines' spool-up time.

There was a considered possibility that the take-off roll type could affect the characteristics of a directivity pattern. As a result, the tail number loggers noted the type of take-off roll for each event.

4.6 Test Director

4.6.1 Deployment

The position of the Test Director was chosen to be in the vicinity of the end of Runway 30 where a potential event could be spotted and external contamination could be most easily judged. This location was also far enough away from the microphones to avoid external contamination of the data.

Prior to deployment, the Test Director synchronized his watch with the TrueTime GPS time code generator. Communications with field personnel at all systems were facilitated by the Test Director to determine when all systems were deployed and ready for data collection. Once the Test Director acknowledged all systems are ready, a 2-way radio broadcast was sent to field personnel informing them that all systems are "go" and measurements were to commence.

4.6.2 During an Event

The Test Director announced twice to all field teams, via 2-way radio, the start of an event and the event number. The Test Director then watched and listened for external noise contamination during the event. The director also monitored the wind speed in real time via the TAMS meteorological system. If the event was deemed "no good" (due to potential external contamination or high wind speeds), the Test Director broadcast this message to all field personnel. The Test Director also noted in the log sheet (included in Appendix B), the event ID, start and end times, average wind speed, idling aircraft in close proximity to the event aircraft, and any comments on external noise.

Figure 23 below summarizes the task sequence during an event.

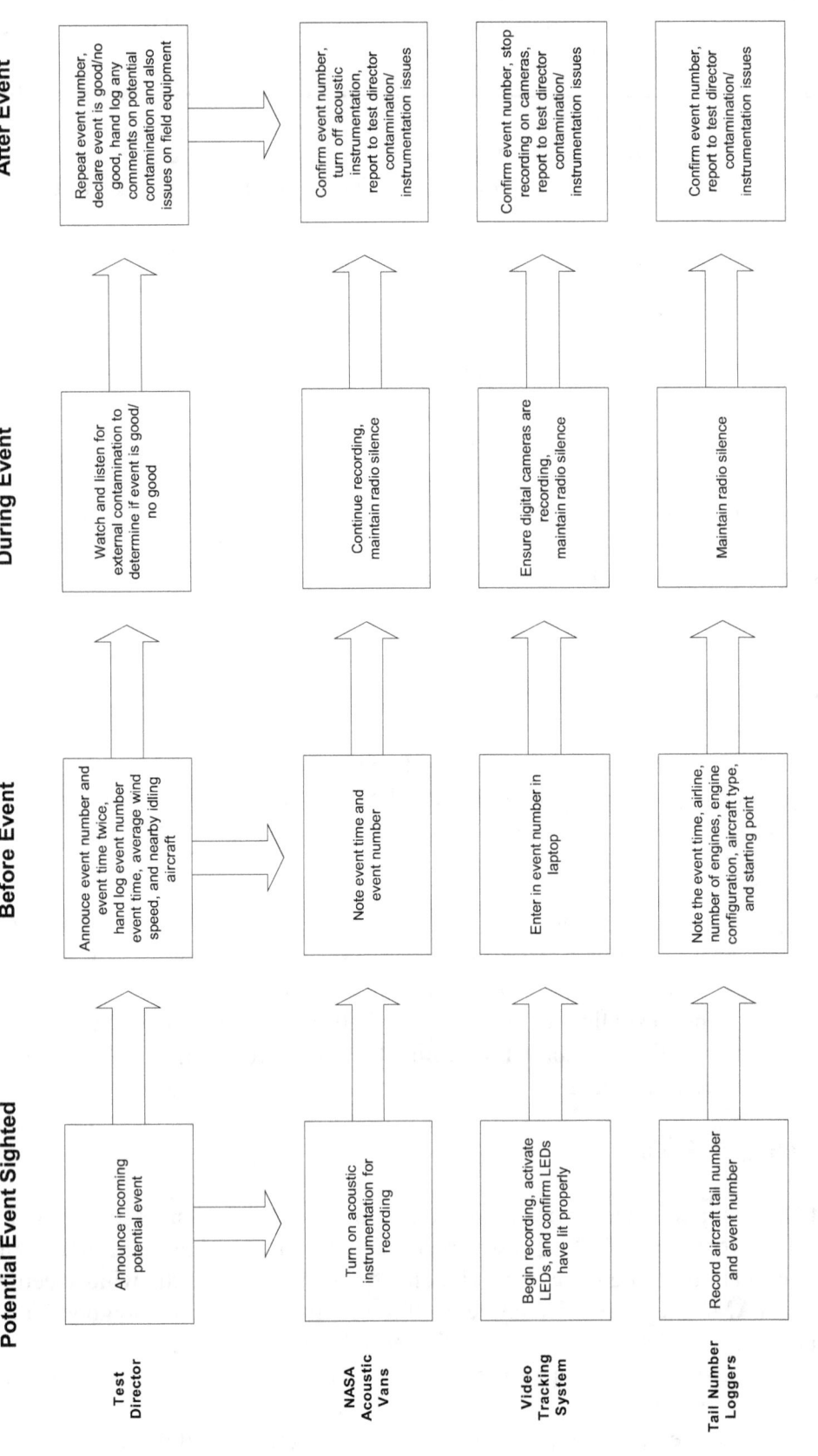

Figure 23. Task Sequence for Each Event

4.7 Communications

Person-to-person communication within individual organizations was facilitated using Nextel 2-way radios. Event information and group broadcasts were facilitated using 2-way CB radios. During an event, there were no radio communications, unless the Test Director deemed the event "no good."

4.8 Quality Assurance

Throughout field measurements, logs were maintained for each system. For example, calibration data were not only recorded by the system, but also manually included in field logs. Field logs also contained information on file names, measurement times on each file, the dates and times of measurements and event numbers. To help supplement the field logs, field photography, diagrams, and maps were used.

Special attention was given to detecting external contamination. Test directors noted any noise sources, e.g., overhead aircraft, construction and airport vehicles that could potentially contaminate an event. Times when the potential contamination event occurred were also noted.

For all instrumentation, raw data were backed up as appropriate at the end of each measurement day to prevent loss of data.

5 DATA REDUCTION AND ANALYSIS

During field measurements data were collected for a total of 508 departure events. Of the 508 events, 390 events were identified as "good events." Good events were those in which field personnel did not observe any potential noise contamination (e.g., noise from overhead aircraft, construction, or airport security vehicles, etc.). The 390 events included a wide variety of aircraft types from turbo-propeller aircraft to four-engine, wide body aircraft. A summary of the events is presented in Tables 4 through 6.

Table 4. Totals for Recorded Departure Events

Aircraft Model	No. of Events	No. of Engines	Wing / Fuselage Mounted	Engine Type
Airbus A310	1	2	Wing	Turbo-Jet
Airbus A319	20	2	Wing	Turbo-Jet
Airbus A320	48	2	Wing	Turbo-Jet
Airbus A330	4	2	Wing	Turbo-Jet
Airbus A340	5	4	Wing	Turbo-Jet
Boeing 717	3	2	Fuselage	Turbo-Jet
Boeing 737	22	2	Wing	Turbo-Jet
Boeing 747	7	4	Wing	Turbo-Jet
Boeing 757	14	2	Wing	Turbo-Jet
Boeing 767	27	2	Wing	Turbo-Jet
Boeing 777	14	2	Wing	Turbo-Jet
Bombardier CL-600	108	2	Fuselage	Turbo-Jet
McDonnell Douglas DC-9	25	2	Fuselage	Turbo-Jet
Turbo-Props	14	2	Wing	Turbo-Prop
Other	78	varies	varies	varies

Total: 390

Table 5. Totals for Turbo-Propeller Aircraft Events

Aircraft Model	No. of Events	No. of Engines	Wing / Fuselage Mounted	Engine Type
Beech B1900D	1	2	Wing	Turbo-Prop
Cessna 425	1	2	Wing	Turbo-Prop
Rockwell 690B	2	2	Wing	Turbo-Prop
Saab-Scania SF340A	10	2	Wing	Turbo-Prop

Total: 14

Table 6. Totals for Aircraft in the "Other" Category

Aircraft Model	No. of Events	No. of Engines	Wing-mounted / Fuselage mounted	Engine Type
Hildebrand Glaster	1	1	Nose	Reciprocating
P Z L Koliber 150A	1	1	Nose	Reciprocating
Cessna 310Q	1	2	Wing	Reciprocating
Piper PA-31-235 Navajo	1	2	Wing	Reciprocating
Piper PA-31-350	1	2	Wing	Reciprocating
BAE 125 SERIES 800A	4	2	Fuselage	Turbo-Jet
BAC 1-11 419/EP	1	2	Fuselage	Turbo-Jet
Cessna 560XL	1	2	Fuselage	Turbo-Jet
Cessna 560	1	2	Fuselage	Turbo-Jet
Cessna 750	3	2	Fuselage	Turbo-Jet
Dassault FALCON 20	1	2	Fuselage	Turbo-Jet
Dassault FALCON 2000	1	2	Fuselage	Turbo-Jet
Embraer EMB-135	1	2	Fuselage	Turbo-Jet
Embraer EMB-135KL	1	2	Fuselage	Turbo-Jet
Embraer EMB-145	2	2	Fuselage	Turbo-Jet
Embraer EMB-145LR	19	2	Fuselage	Turbo-Jet
Embraer EMB-145XR	1	2	Fuselage	Turbo-Jet
Grumman G-1159	1	2	Fuselage	Turbo-Jet
Gulfstream G-1159A	2	2	Fuselage	Turbo-Jet
Gulfstream G-IV	5	2	Fuselage	Turbo-Jet
Gulfstream G-V	1	2	Fuselage	Turbo-Jet
Hawker Siddely 125 SERIES 700A	1	2	Fuselage	Turbo-Jet
Israel Aircraft 1124A	1	2	Fuselage	Turbo-Jet
Israel Aircraft G-200	1	2	Fuselage	Turbo-Jet
Lear Jet 24D	1	2	Fuselage	Turbo-Jet
Lear Jet 45	1	2	Fuselage	Turbo-Jet
Lear Jet 55	1	2	Fuselage	Turbo-Jet
Lear Jet 60	2	2	Fuselage	Turbo-Jet
McDonnel Douglas MD-88	4	2	Fuselage	Turbo-Jet
Raytheon 400A	2	2	Fuselage	Turbo-Jet
Dassault FALCON 900 EX	1	3	Fuselage	Turbo-Jet
Dassault FALCON 900B	3	3	Fuselage	Turbo-Jet
Dassault MYSTERE-FALCON 50	3	3	Fuselage	Turbo-Jet
Vickers VC-10	1	4	Fuselage	Turbo-Jet
DORNIER 328-300	4	2	Wing	Turbo-Jet
Embraer ERJ 170-100	1	2	Wing	Turbo-Jet
BAE 146 SERIES 200A	1	4	Wing	Turbo-Jet
Total:	**78**			

5.1 Data Reduction Process

Continuous, one-third octave-band acoustic data, aircraft position information, tail numbers, and meteorological data were collected and processed. This section discusses data reduction and aggregation.

5.1.1 Acoustic Data

Acoustic data collected by the NASA acoustic vans and portable acoustic measurement systems were processed using a Volpe-developed software program, Noiselogger™ Data Analysis Tool (NDAT). NDAT is a Windows-based program that uses one-third octave spectral time history data and event times as input and plots the overall sound pressure levels for each event. The tool allows the user to visually inspect the recorded data for contamination not detected by field personnel.

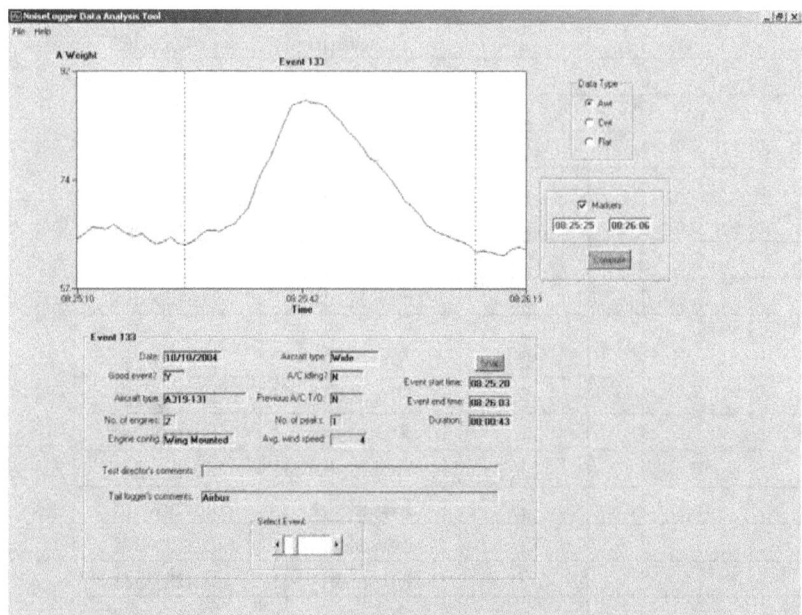

Figure 24. NDAT Software Program

The program also imports and displays all pertinent event information recorded in the field, including aircraft type, number of engines, and Test Director's comments to aid the user. Once the user has visually inspected the data for an event and has fine-tuned the event duration for metric computation, the program computes Sound Exposure Levels (L_E) and Equivalent Sound Levels (L_{eq}) for multiple averaging times. The program is capable of computing results with multiple weighting options (A-weighting, C-weighting, Z-weighting -- or no weighting). The majority of aircraft noise prediction models based on SAE-AIR-1845 use the A-weighted Sound Exposure Level (L_{AE}) metric. As a result,

L_{AE} was the primary metric used for this study. An analysis (discussed in Section 5.3) later indicated there is little sensitivity to the use of either L_{Aeq} or L_{AE}.

5.1.2 Aircraft Tail Numbers

Aircraft tail numbers bridged the relationship between the measured acoustic data and specific aircraft type. Tail numbers obtained from field loggers were entered into an aircraft registration website[9] to identify the aircraft type for each event. Comments noted by the primary and secondary tail loggers such as aircraft type, number of engines, engine configuration, and best guess of aircraft model were compared to results from the registration website to ensure the proper aircraft was identified.

5.1.3 Meteorological Data

Meteorological data obtained during the study characterized the wind speed and direction at the measurement site. Aside from a single day where measurements were postponed because of high winds, wind conditions during the measurement period were relatively calm. As noted in Section 4.4.2, AIR-1845 recommends that data should not be collected when wind speeds exceed 15 knots (approximately 18 mph). Test Directors monitoring the TAMS meteorological unit did not detect wind exceeding 18 mph during measurements of any of the events. An analysis of the meteorological data confirmed that at no time during the measurement period did wind speeds exceed 13.3 mph. Figure 25 displays the highest instantaneous wind speed recorded each day during measurements.

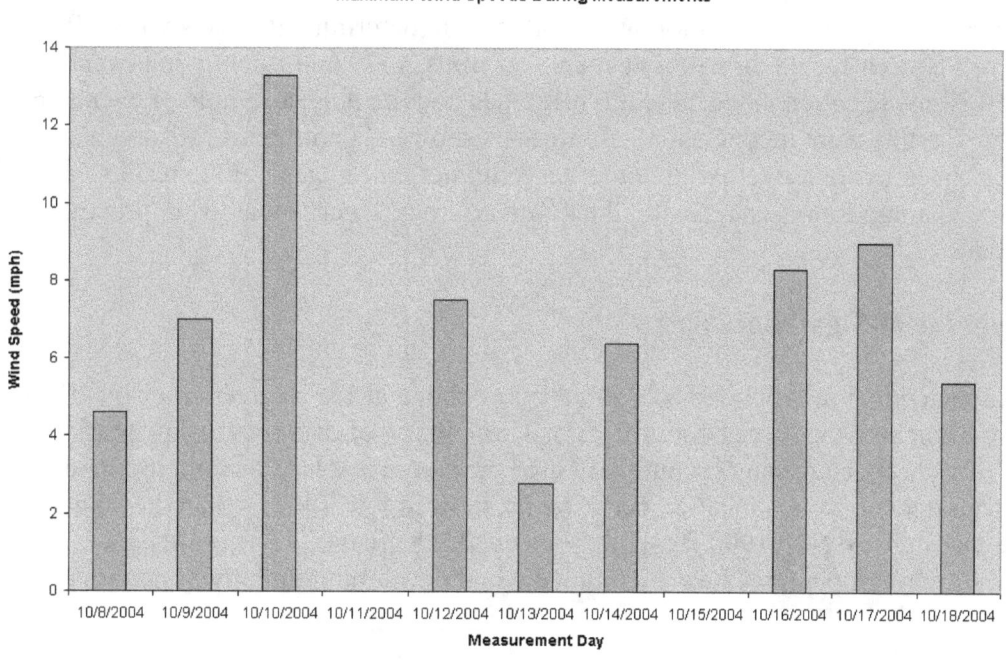

Figure 25. Maximum Wind Speeds during Measurement Days

5.1.4 Quality Assurance

The quality of the measured data is crucial since the data will be used in development of the updated AIR-1845 directivity algorithm as well as in future versions of INM/AEDT. Aircraft tail numbers recorded by the primary and secondary tail number loggers were compared to ensure the proper aircraft were identified. Special care was given to inspecting the acoustic data during processing to ensure no external contamination was included during calculations of L_{AE} and L_{Aeq}. For quality assurance purposes, the following data were not included in final analysis:

- Data from events for which contamination was evident during visual inspection of the event's sound level time history
- Data from events for which aircraft departed after turning from taxiway "Z" (i.e., the taxiway approximately 600 ft west of the most commonly used taxiway, "Q")
- Unavailable data during the malfunction of microphones, as identified by field personnel
- Data collected from microphone N2 (see Section 5.1.5.3)
- Data for aircraft in the "Other" category since there was either not a sufficient (i.e., representative) pool of events for each aircraft type or the aircraft type only represented a small part of the 2005 fleet

5.1.4.1 Take-off Roll Type

As discussed in Section 4.5.2, there was a considered possibility that the take-off roll type of an aircraft may affect the directivity pattern. Field personnel noted the type of take-off roll for each event. An investigation was conducted to determine if events with different take-off roll types affected directivity patterns. Airbus A320 and Boeing 767 events were used for this investigation since these aircraft types had the largest sample of events in the Study. Results were inconclusive. Some tests showed "continuous roll" events exhibiting lower overall levels while other tests did not, but in general the results seemed to be somewhat random. As a result, all take-off roll types were included in processing of final data.

5.1.4.2 Line-of-Sight Blockage

It was suspected that microphone N2 (900 ft microphone at 135 degrees) may be affected by line-of-sight blockage (see Section 3.1.2). Completion of data processing confirmed that line of sight blockage significantly affected measured levels at this microphone. Overall sound levels at this location consistently showed lower levels than the adjacent, alternate microphone N2A (900 ft microphone at 135 degrees). As a result, data collected from microphone N2 were excluded when developing directivity patterns.

5.1.4.3 Idling Aircraft and Terminal Noise

Individual events where field personnel observed idling aircraft in the queue were compared to known acoustically clean events (i.e., those without idling aircraft). The Airbus A320 and Boeing 767 events were again selected for this investigation. The resulting comparisons did not show conclusive evidence that data collected was affected by idling aircraft in the queue.

A preliminary investigation was also undertaken on individual events at microphone N1 (900 ft microphone at 120 degrees) to investigate why there are often higher sound levels at this position, as compared to V1 (900 ft microphone at 240 degrees). N1 and V1 are symmetrically located with respect to the runway centerline (Figure 26). A comparison of L_{Aeq} spectral data was performed for microphones N1 and V1 in an attempt to quantify any potential effects. Initial results showed that N1 was exposed to more sound level energy at certain frequencies. This effect may be due to the fact that microphone N1 is physically located in closer proximity to noise emanating from the airport terminal, but further investigation is warranted.

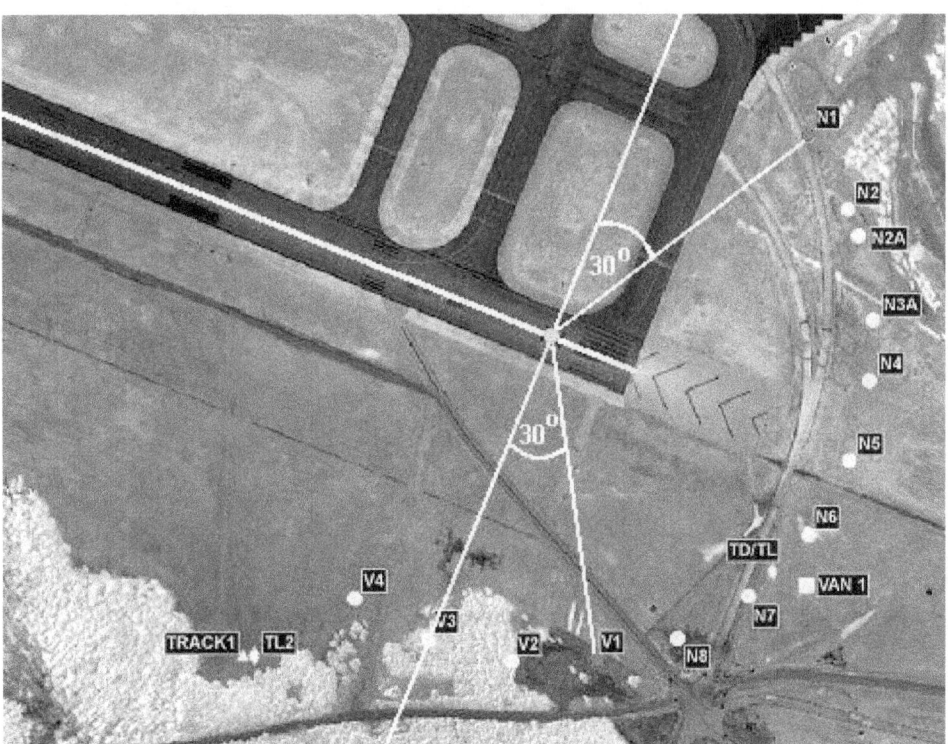

Figure 26. Symmetrical Microphones N1 and V1 at 900 ft

5.1.4.4 Data Collected at 1950 ft

Data collected at 1950 ft were not compared to AIR-1845 but are included in Appendix E of this report. This data were not included in the detailed final analysis because of insufficient signal to noise ratio due to: 1) Microphones at 1950 ft were partially located in the vicinity of building construction during measurements; and 2) Microphones V5 and

V6 were located in close proximity to the airport terminal. Because symmetrical microphones were not present at 1950 ft as they were at 900 ft, data were unavailable for further investigation.

5.2 Deriving Directivity Patterns

Directivity patterns were created for each aircraft model measured during data collection. The following steps were used to derive the directivity patterns:

1) Calculated L_{AE} for data recorded at each microphone using the methodology explained in Section 5.1.1;
2) Grouped resulting L_{AE} by aircraft type;
3) For individual aircraft types, calculated the arithmetic mean L_{AE} for events at each microphone; and
4) For individual aircraft type, plotted the arithmetic mean L_{AE} at each microphone location on polar plots as shown in the example Figure 27.

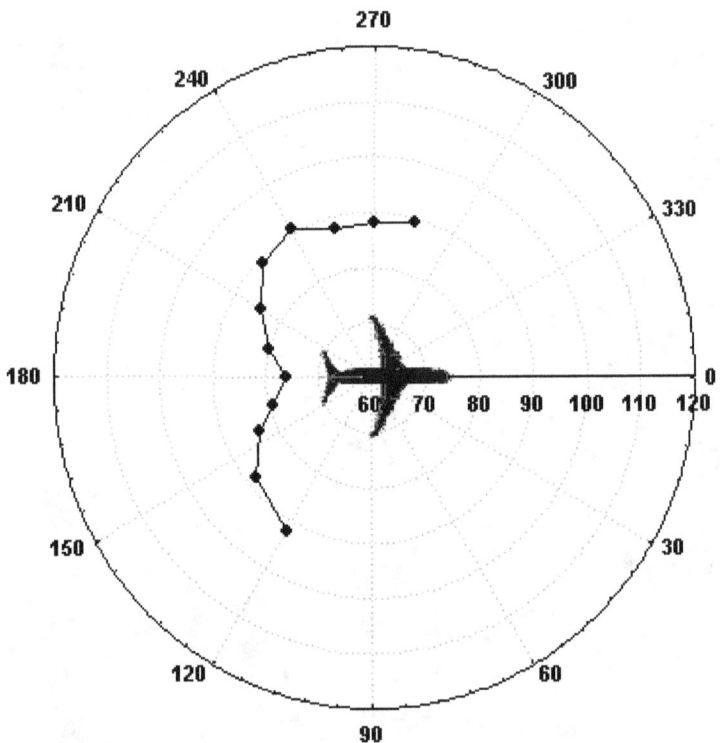

Figure 27. Mean L_{AE} at 900 ft Microphone Locations for the Airbus A319

5.2.1 Noise Metrics

During processing of the acoustic data, several different metrics were derived, including A-weighted equivalent sound level L_{Aeq} (for multiple averaging times), A-weighted Sound Exposure Level (SEL, denoted by L_{AE}) and un-weighted exposure level (L_{ZE}). An analysis was undertaken to compare data processed using the different metrics. Example results, displayed in Figures 28 and 29, indicate there is little sensitivity to the use of either L_{Aeq} or L_{AE}. This may also suggest that the level at L_{ASmx} predominantly defines the directivity pattern. Analysis using the un-weighted exposure level showed more of a pronounced indent at 180 degrees.

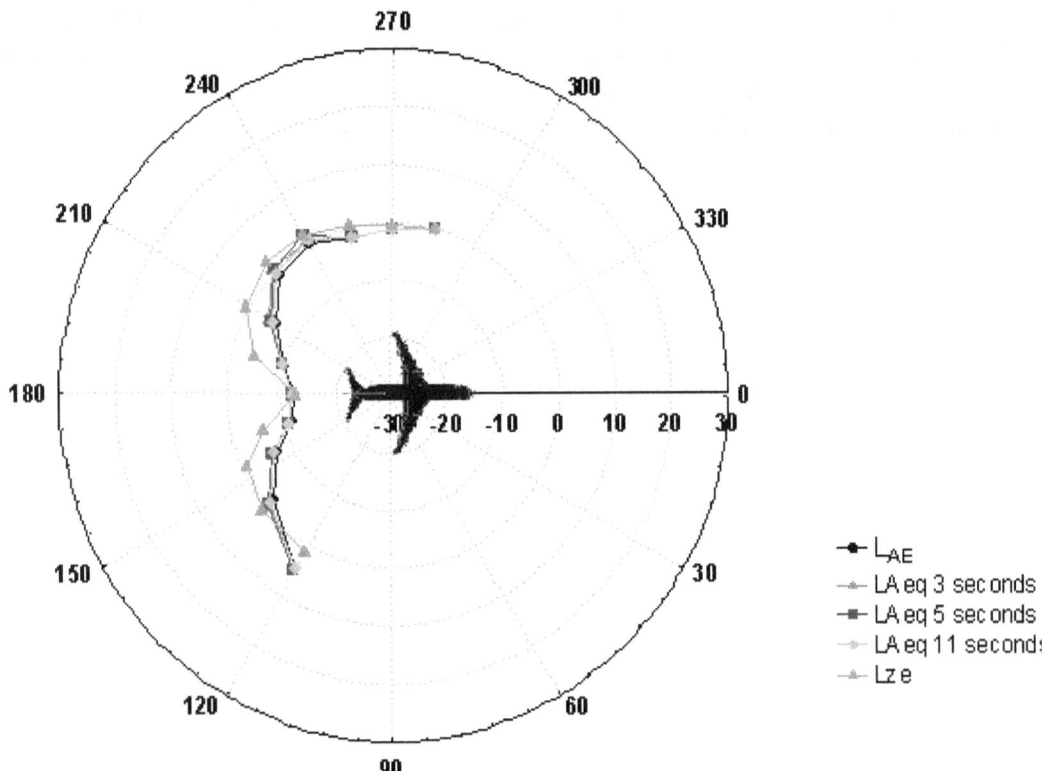

Figure 28. Comparison between Different Metrics Used to Process A320 Events

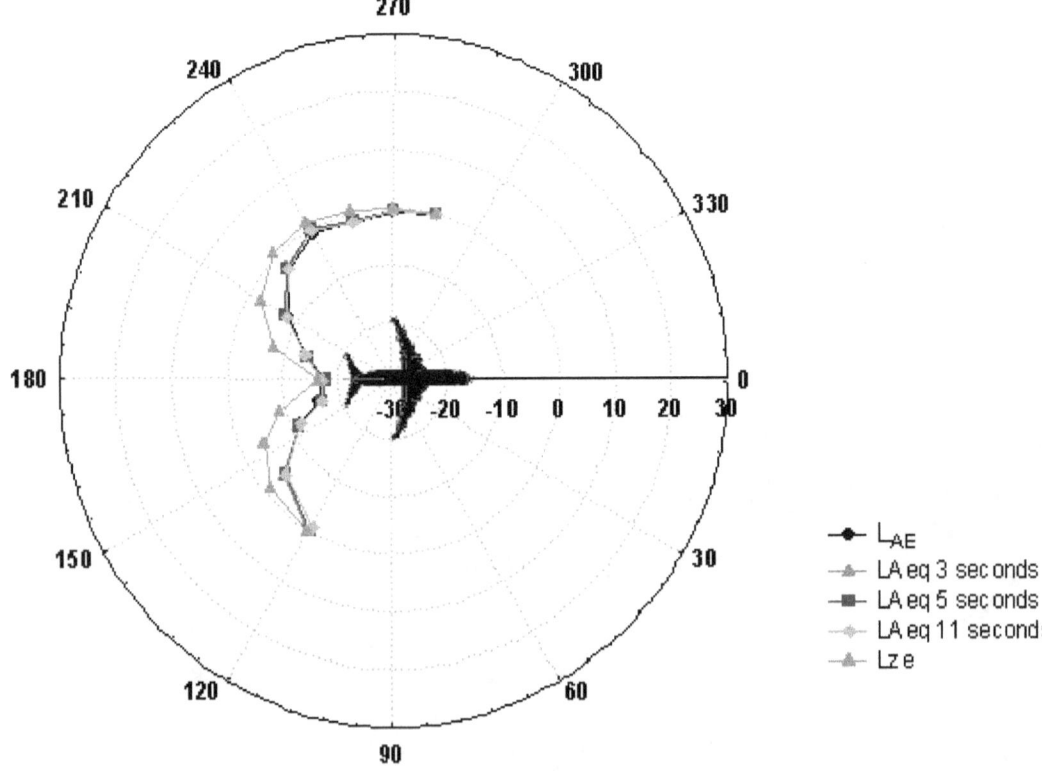

Figure 29. Comparison between Different Metrics Used to Process B767 Events

5.2.2 Normalizing Data to 285-Degrees

The current AIR-1845 directivity pattern (shown in Figure 30) is presented as "corrections" as opposed to absolute levels.

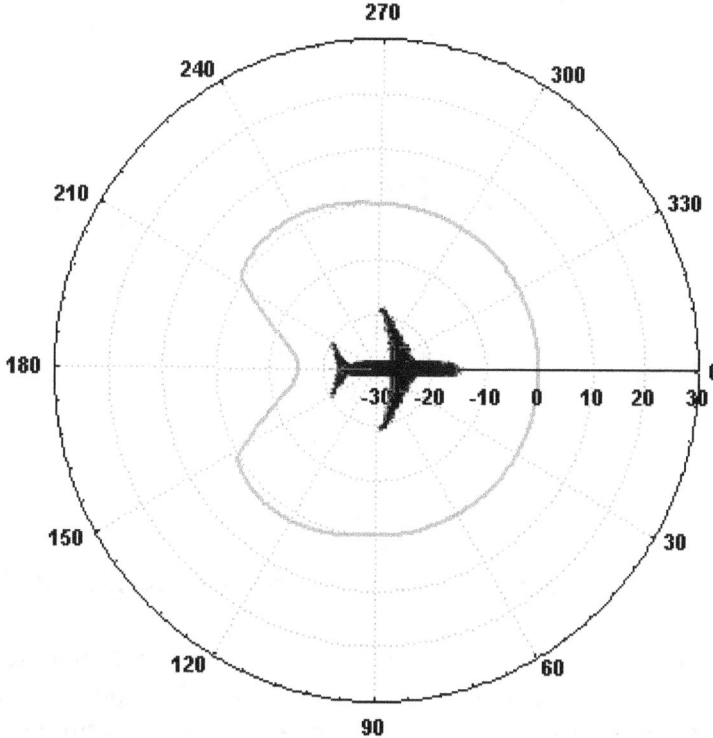

Figure 30. AIR-1845 Directivity Pattern at 900 ft

To allow comparison with the AIR-1845 algorithm, the computed directivity patterns had to be converted into directivity corrections. This conversion was accomplished by using the L_{AE} data from the current study and normalizing it, by subtracting the level at the 285-degree microphone from the levels at all other microphones. Since the 1980 algorithm prescribes a 0 dB correction at 285 degrees, normalizing the current data to the L_{AE} at 285-degree level readily allows for comparison of the new data with the older algorithm. Figure 31 shows an example of converting the directivity pattern from Figure 27 to directivity corrections. Note that normalizing absolute L_{AE} to the 285-degree microphone did not change the shape of the pattern.

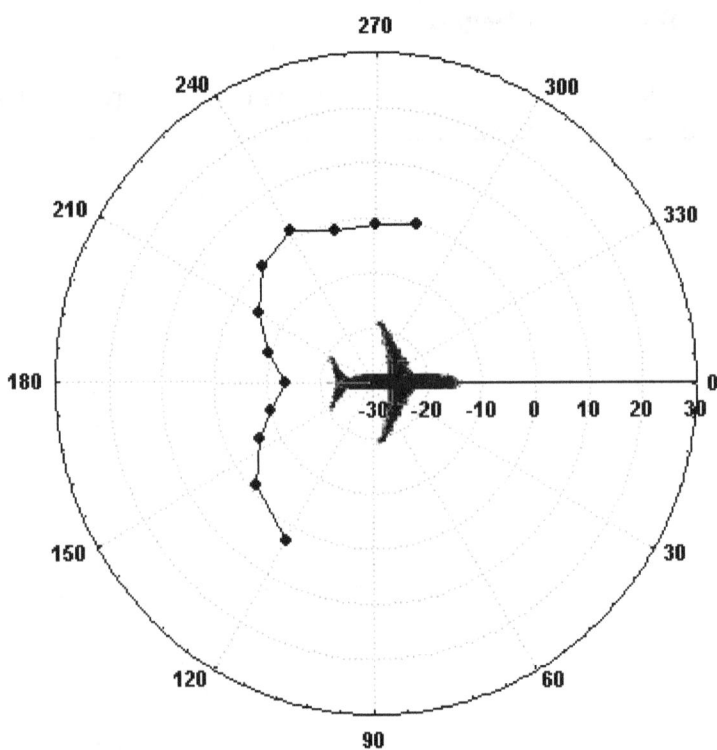

Figure 31. Directivity Pattern for the Airbus A319 at 900 ft

To confirm the reasonableness of the 0 dB correction at the 285-degree microphone, an analysis was performed to compare the measured versus INM-modeled values at that angle. Take-off weight, thrust configurations, aircraft speed and altitude were obtained from the Aircraft Communication Addressing and Reporting System (ACARS) of a major U.S. carrier operating at IAD during the study. ACARS is a digital data link transmitted from the aircraft via VHF radio to individual airline archives. Temperature data were obtained from the TAMS meteorological station. These data were then entered into INM and computed for a variety of aircraft where ACARS data were available. The resultant comparisons are presented in Table 7. The analysis shows overall, good agreement between measured versus modeled results at 285-degrees.

Table 7. Measured vs. INM-Modeled Comparison for a Microphone at 285-Degrees

Event	Aircraft Type	Measured	Modeled	Delta
164	A319	88.3	89.1	-0.8
271	A320	88.3	91.1	-2.8
207	A320	88.8	91.0	-2.2
303	A320	88.9	92.0	-3.1
306	A320	89.1	92.0	-2.9
372	B737	87.0	88.1	-1.1
450	B747	96.3	97.4	-1.1
277	B757	89.9	91.3	-1.4
208	B757	90.7	91.3	-0.6
225	B767	95.0	97.0	-2.0
	Average:	90.2	92.0	-1.8

5.3 Comparison of Newly Developed Patterns with AIR-1845

An analysis was conducted to determine if the newly developed directivity patterns were different from the AIR-1845 directivity patterns. T-tests were utilized to determine if the data measured at each angular offset position were statistically different from the SAE-1845 adjustment at that position. These tests answered the question: "Is there at least a 95% probability that the mean of the data is statistically similar to the value of the AIR-1845 adjustment?"

The average measured corrections and corresponding +-95% CI are plotted along with the AIR-1845 directivity pattern (displayed as a continuous, gray line) and shown in Figures 34-46. Points which are statistically similar to the reference are shown in blue, points are not statistically similar to the reference are shown in red.

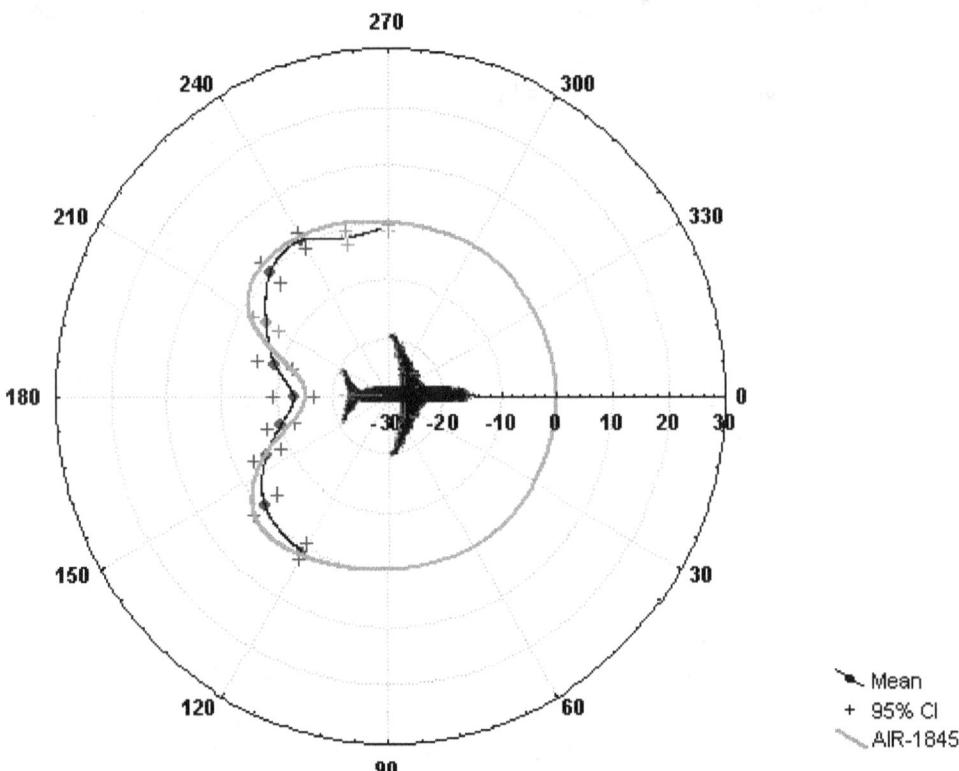

Figure 32. Airbus A319 Directivity Pattern at 900 ft (16 Events)

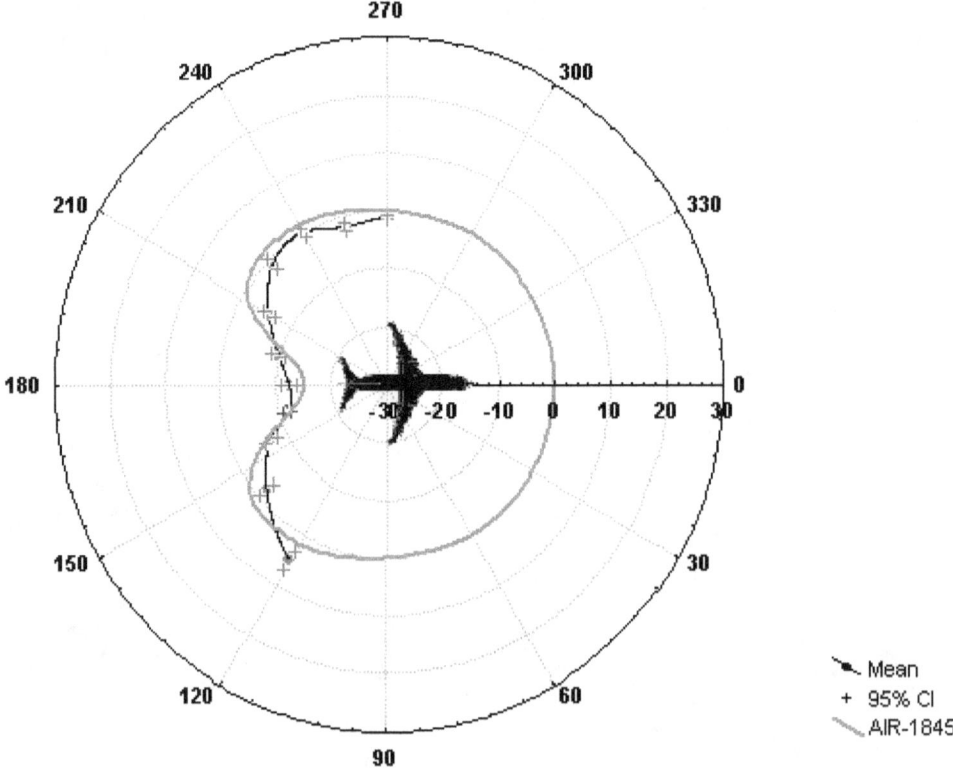

Figure 33. Airbus A320 Directivity Pattern at 900 ft (48 Events)

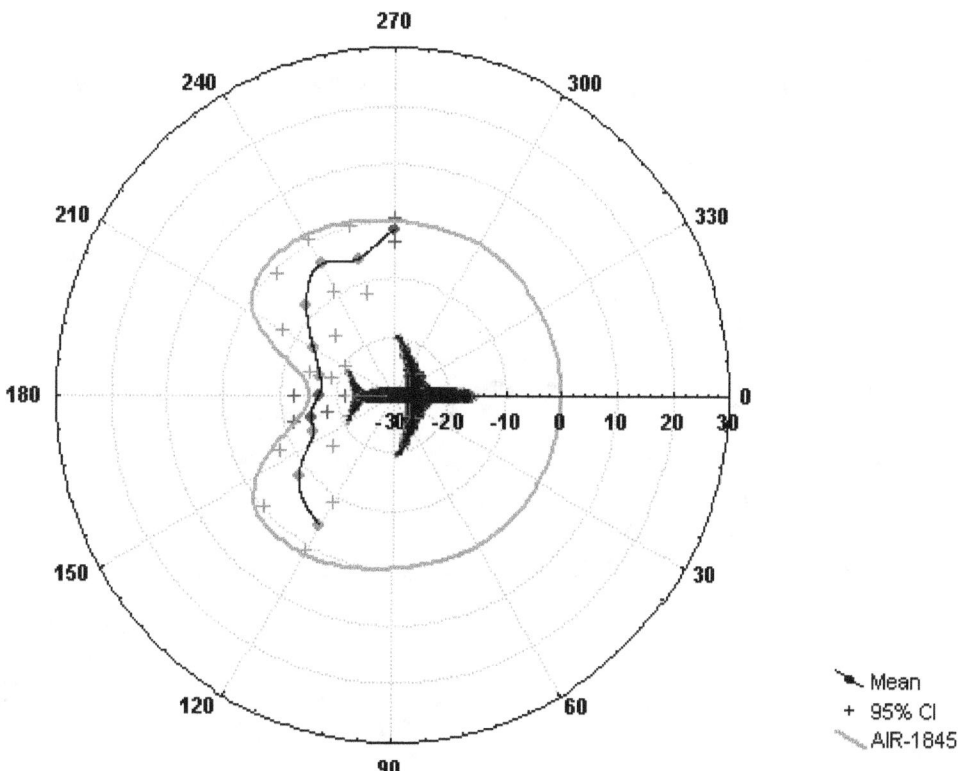

Figure 34. Airbus A330 Directivity Pattern at 900 ft (4 Events)

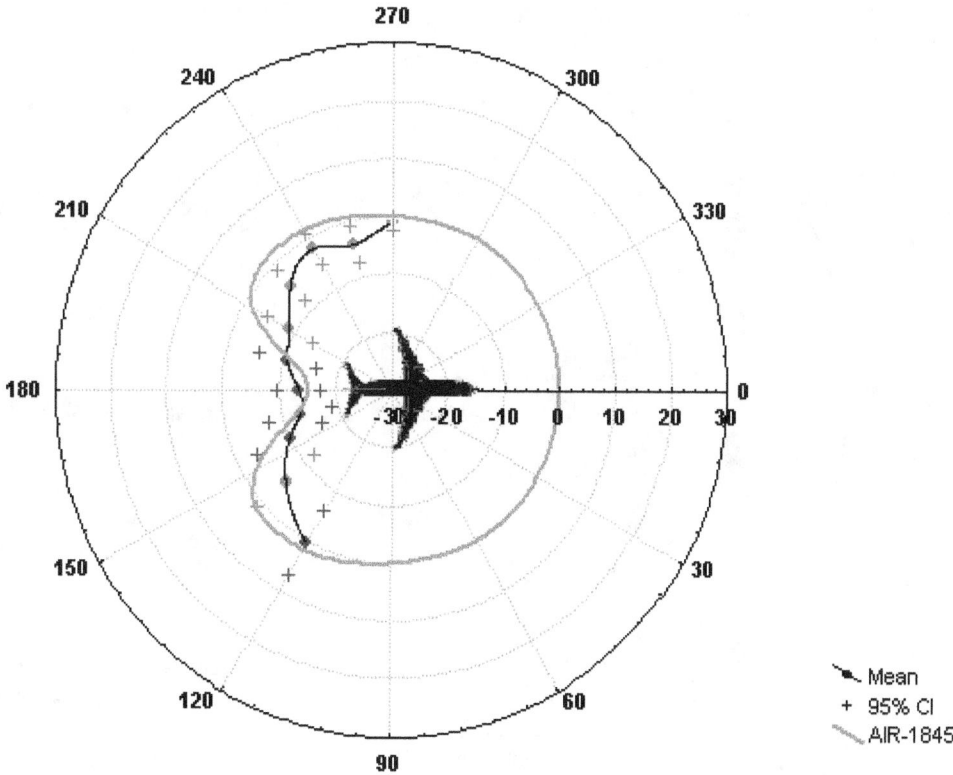

Figure 35. Airbus A340 Directivity Pattern at 900 ft (5 Events)

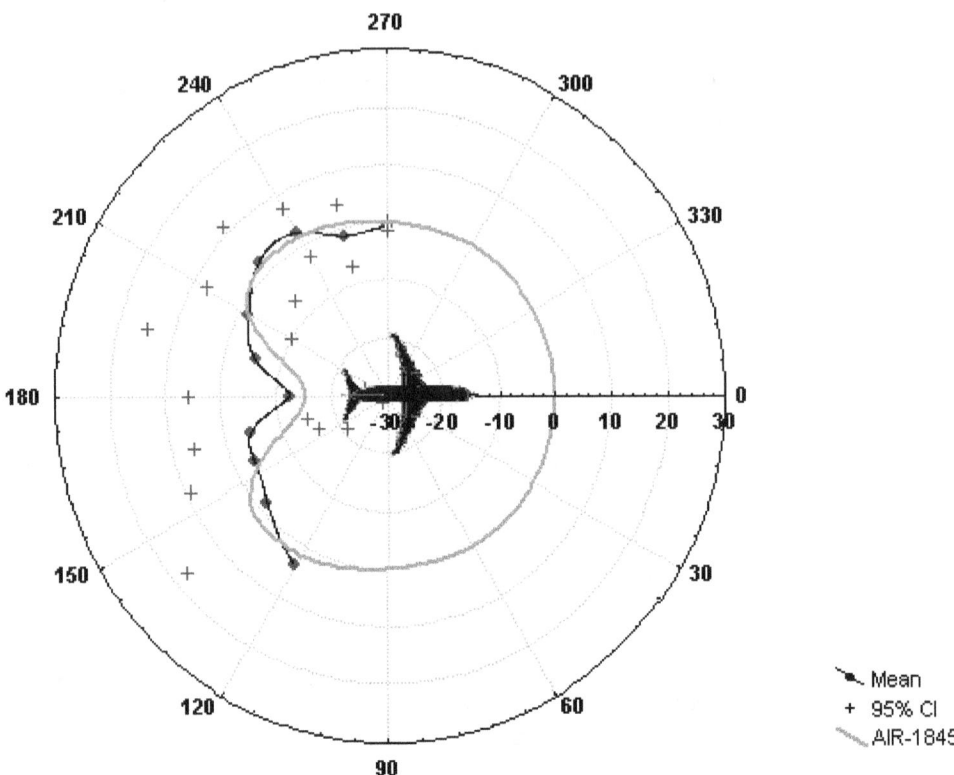

Figure 36. Boeing 717 Directivity Pattern at 900 ft (3 Events)

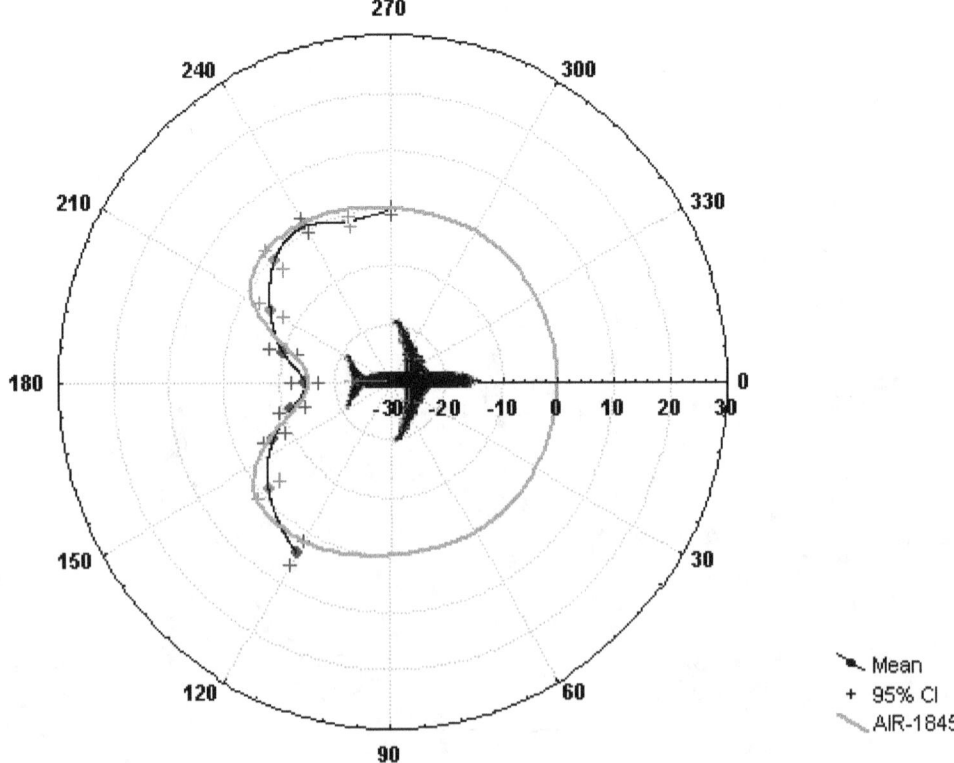

Figure 37. Boeing 737 Directivity Pattern at 900 ft (20 Events)

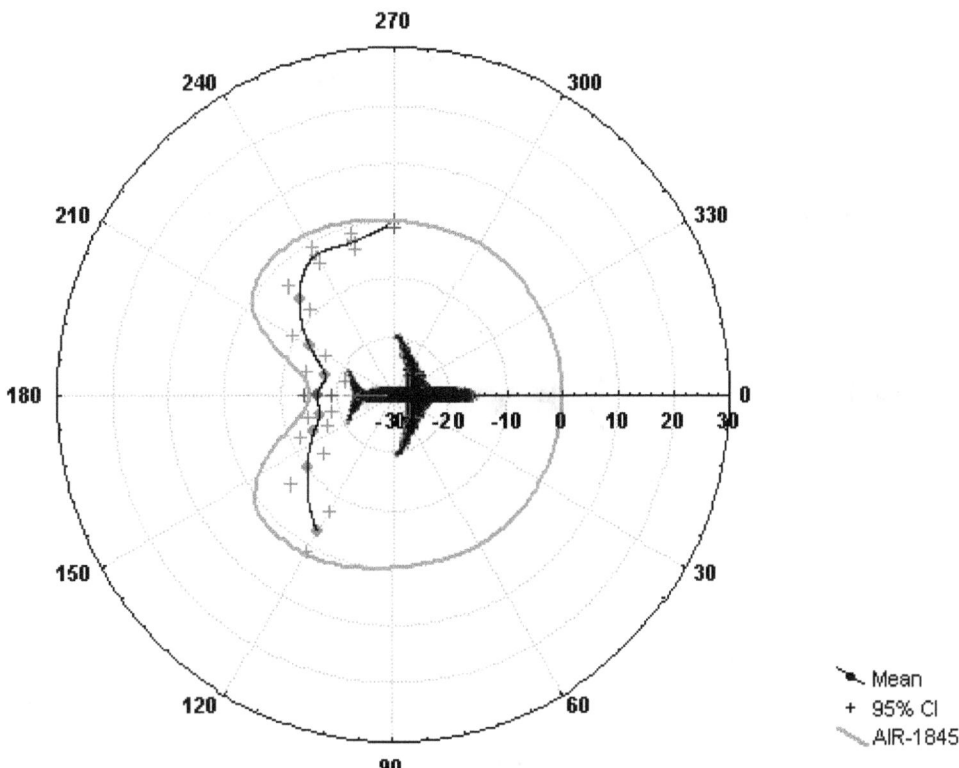

Figure 38. Boeing 747 Directivity Pattern at 900 ft (7 Events)

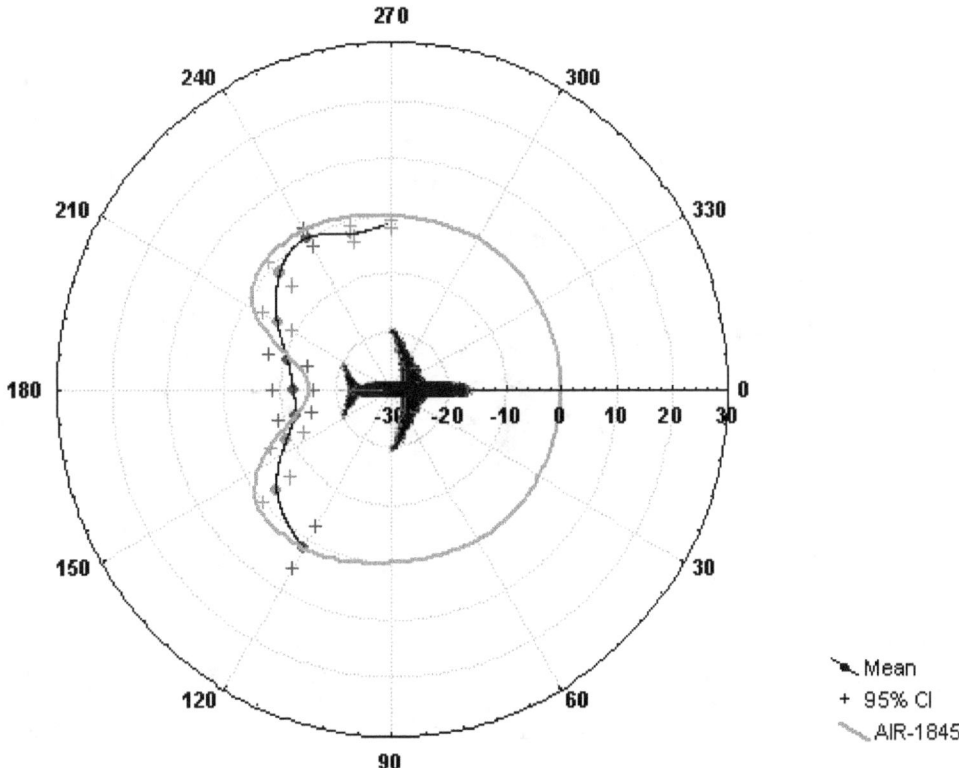

Figure 39. Boeing 757 Directivity Pattern at 900 ft (12 Events)

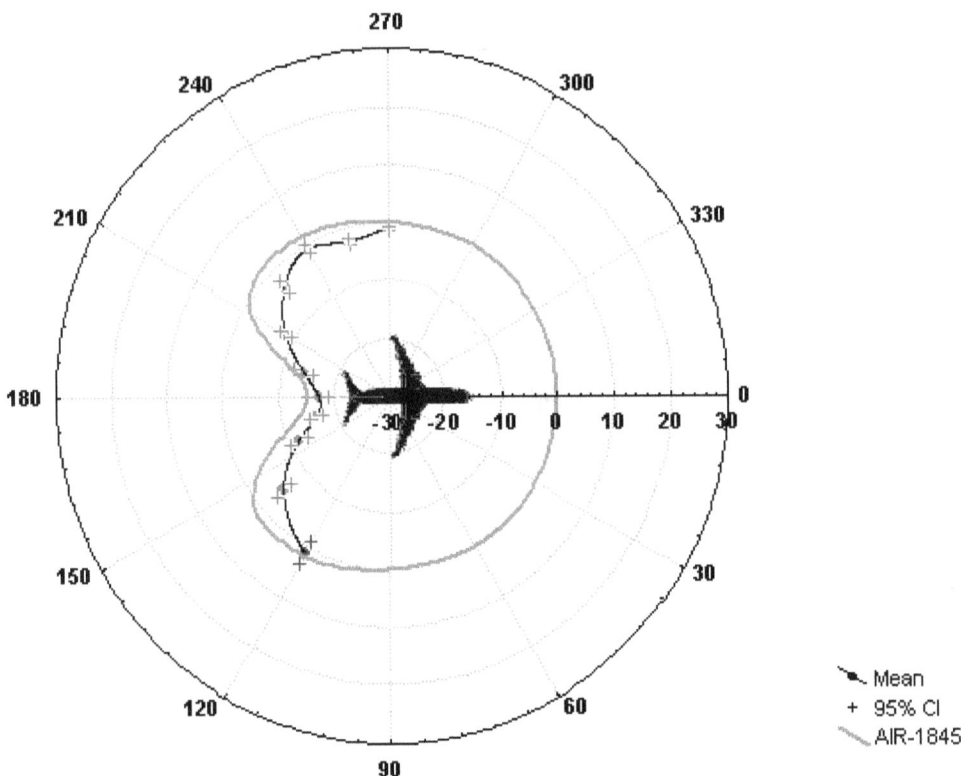

Figure 40. Boeing 767 Directivity Pattern at 900 ft (27 Events)

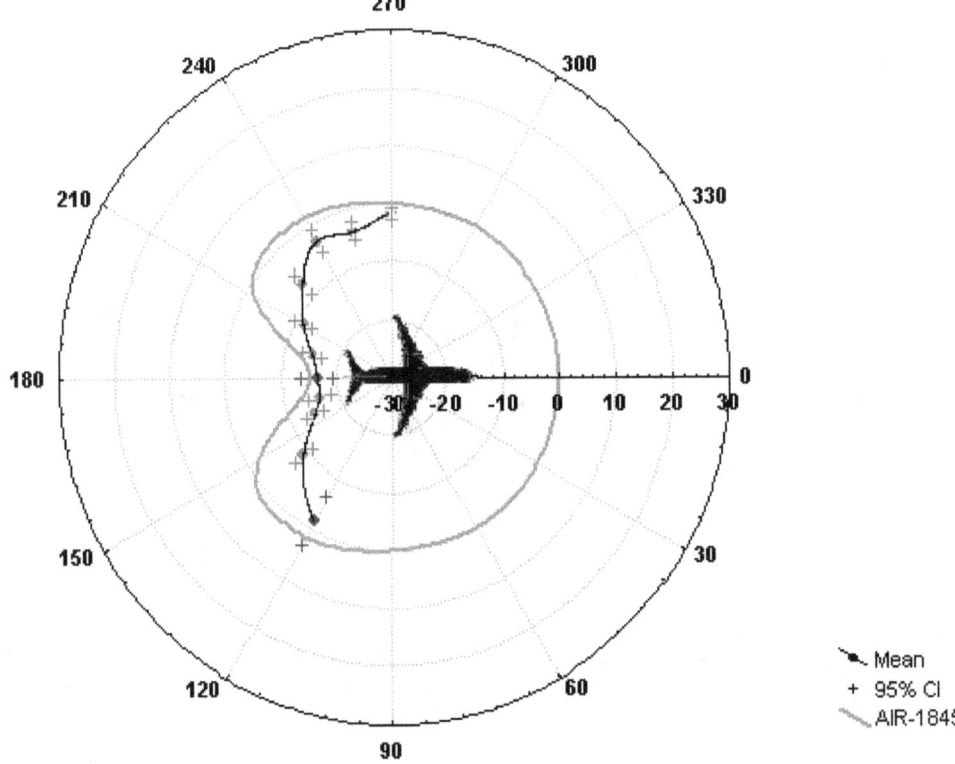

Figure 41. Boeing 777 Directivity Pattern at 900 ft (12 Events)

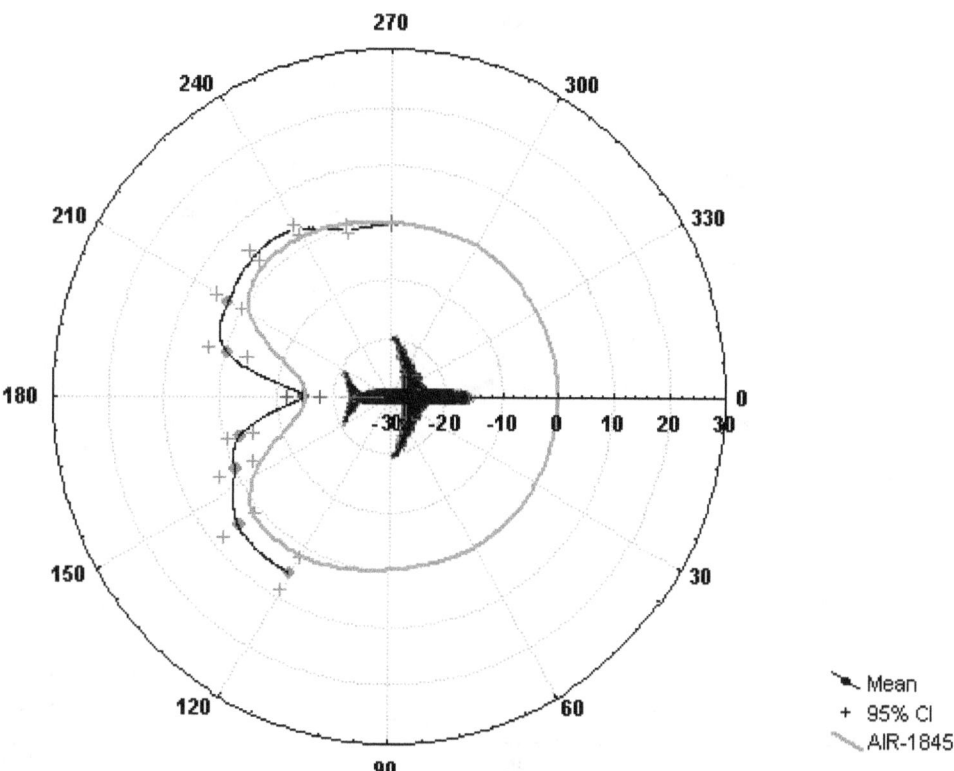

Figure 42. McDonnell Douglas DC9 Directivity Pattern at 900 ft (23 Events)

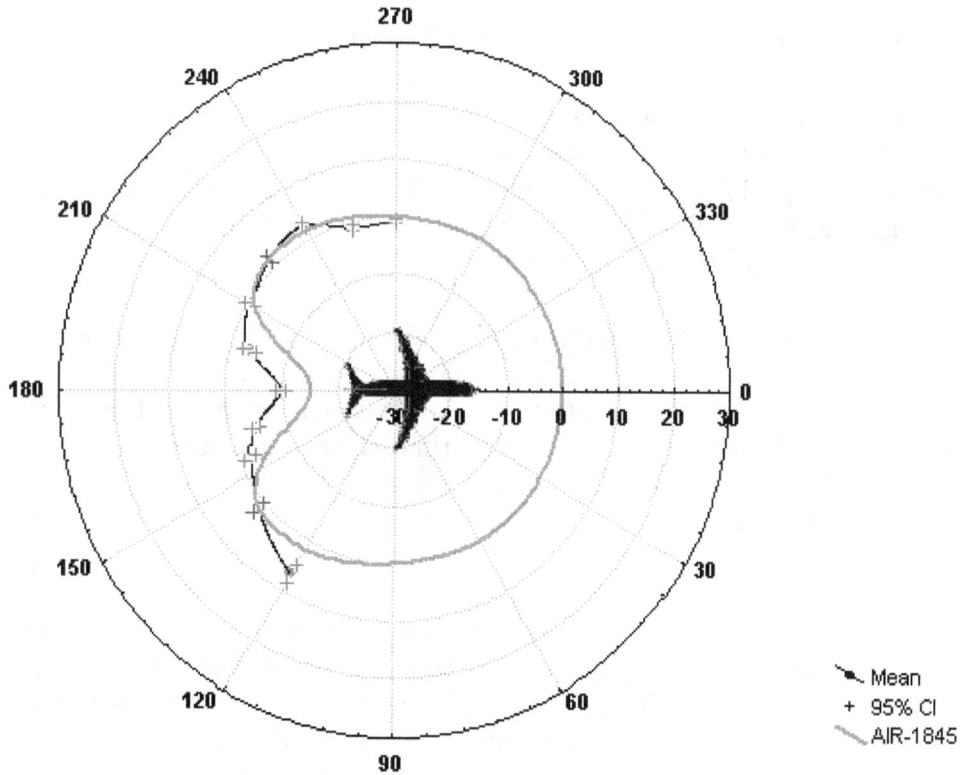

Figure 43. Bombardier CL600 Directivity Pattern at 900 ft (104 Events)

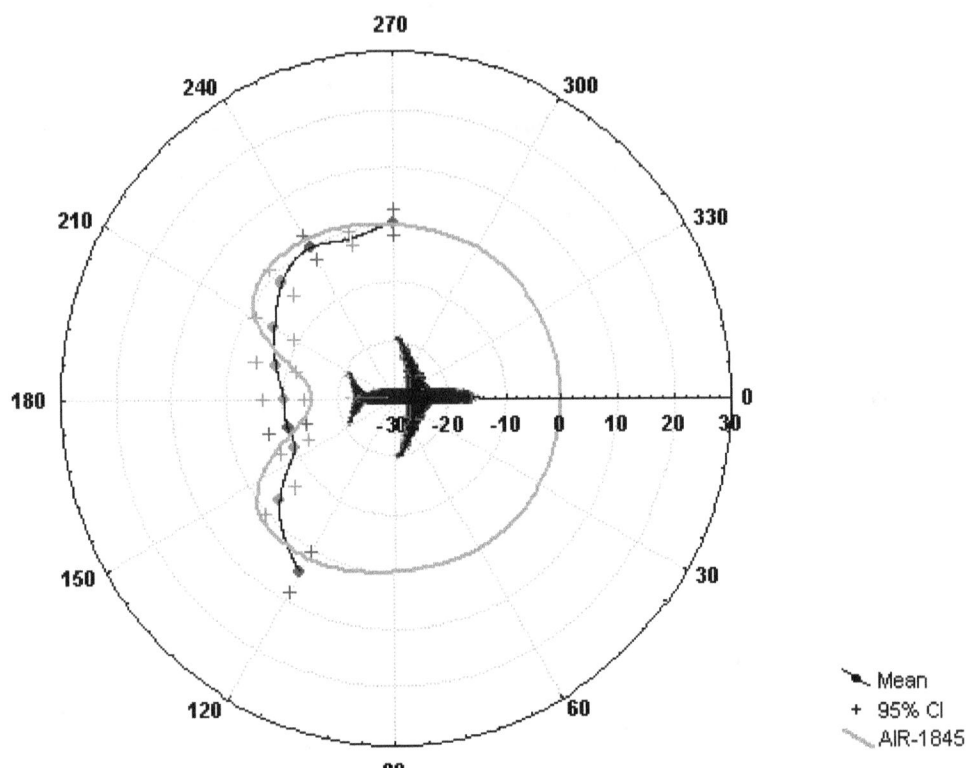

Figure 44. Turbo-Propeller Aircraft Directivity Pattern at 900 ft (14 Events)

An analysis of the comparisons concluded the following:

- The majority of jet aircraft show similarity with the current standard around 180 degrees, but differ in the "lobes" from 120-150 and 210-240 degrees
- The Boeing 717 is the only aircraft, which exhibits statistical similarity at all polar angles with the AIR-1845 algorithm. This, however, is a result of the fact that there were only 3 events for this aircraft

5.4 Proposed Updated Directivity Adjustment

The proposed, updated SOTR directivity adjustments were computed from the data through a fleet-weighted average. First, the arithmetic average noise level was computed for each aircraft type at each azimuth angle from 90 to 180 degrees (symmetrical directivity was assumed). Then the SOTR directivity noise levels were normalized to 0.0 dB at 285 degrees.

Through the comparison of the average, aircraft-specific SOTR directivities, it became apparent that a single directivity adjustment would be insufficient in describing the behind SOTR noise levels of today's fleet. Since turboprop's SOTR directivities were significantly different from the directivities of jet aircraft, they were evaluated separately. Two sets of SOTR directivity adjustments were developed; one for jet aircraft, and the other for turboprop aircraft.

For jet aircraft, a weighted-regression was used to compute the SOTR directivity adjustment. The weighting factor was chosen to modify the regression such that aircraft were represented proportionate to their relative operational presence in the 2005 fleet[*]. The individual aircraft types measured represented 85% of the aircraft types in the 2005 fleet. Descriptions of the aircraft fleet used for these derivations are presented in Appendix F of this report.

For turboprop aircraft, a more standard regression fitting process was used to develop the SOTR directivity adjustment. The regression was computed using all data from the events measured. No weighting factor was applied, as the measured data set was smaller (14 events) and could not adequately represent the majority of the fleet of turboprop aircraft.

This resulted in two sets of updated SOTR directivity adjustments. The updated SOTR directivity adjustment for jet aircraft is represented by the following equation:

For $\quad 90° \leq \theta < 180°$

$$DIR_{ADJ} = 2329.44 - (8.0573\theta) + \left(11.51 \exp\left(\frac{\pi\theta}{180}\right)\right) - \left(\frac{3.4601\theta}{\ln\left(\frac{\pi\theta}{180}\right)}\right) - \left(\frac{17403338.3 \ln\left(\frac{\pi\theta}{180}\right)}{(\theta)^2}\right) \quad (3)$$

The SOTR directivity adjustment for turboprop aircraft is represented by the following equation:

For $\quad 90° \leq \theta < 180°$

$$DIR_{ADJ} = -34643.898 + \left(\frac{30722162}{\theta}\right) - \left(\frac{11491573931}{\theta^2}\right) + \left(\frac{2.34929 \times 10^{12}}{\theta^3}\right) - \left(\frac{2.83584 \times 10^{14}}{\theta^4}\right) + \left(\frac{2.02272 \times 10^{16}}{\theta^5}\right) - \left(\frac{7.90084 \times 10^{17}}{\theta^6}\right) + \left(\frac{1.30507 \times 10^{19}}{\theta^7}\right) \quad (4)$$

For both jet and turboprop aircraft, the proposed SOTR directivity adjustments are 0.0 dB for azimuth angles between 0 and 90 degrees, and are symmetrical on either side of the

---

[*] AEDT Common Operations Database (COD) for the FAA Destination 2025 environmental inventory of the calendar year 2005.

aircraft. The smoothing algorithm used at distances greater than 2500 ft for the current SOTR implementation was not used for the proposed implementation.

The proposed and current SOTR directivity adjustments are presented in Figure 45. Furthermore, the differences between the proposed and current SOTR directivity adjustments are presented in Figure 46.

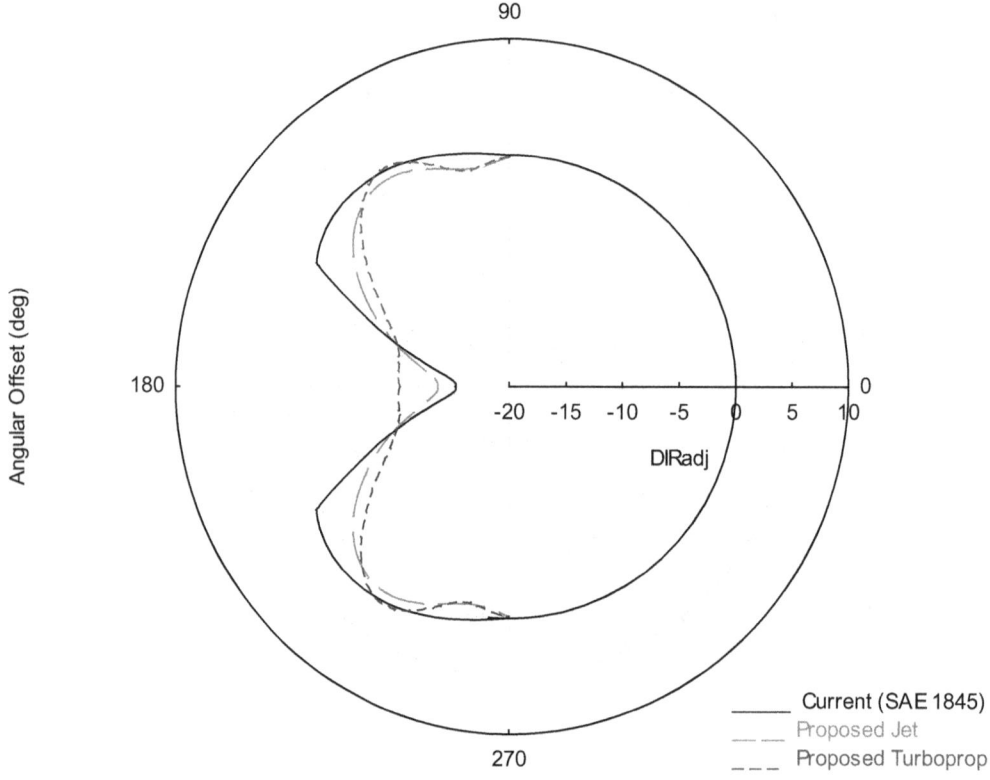

Figure 45. The Current and Proposed behind Start of Take-off Roll Directivity Adjustments

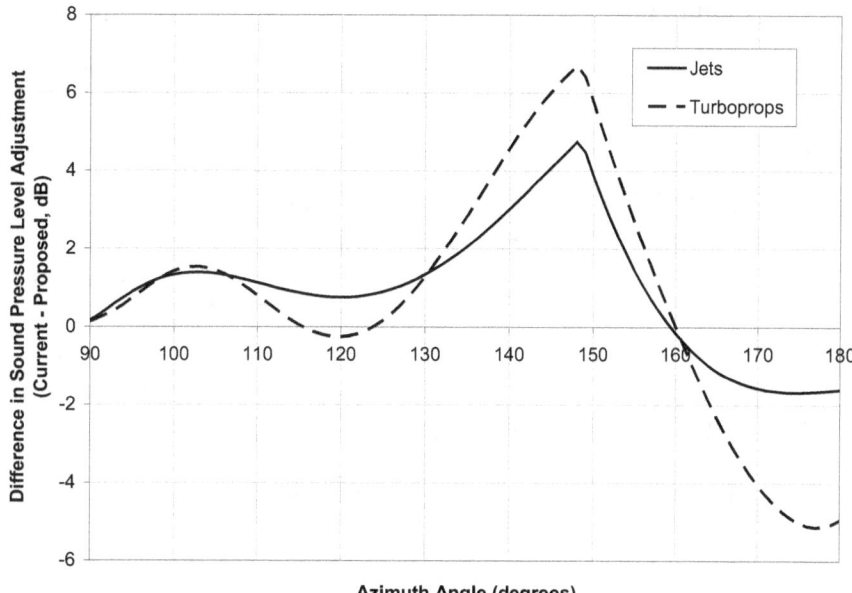

Figure 46. The Differences between the Current and Proposed behind Start of Take-off Roll Directivity Adjustments for Jet and Turboprop Aircraft

5.5 Additional Directivity Investigations

Several alternative SOTR implementations for jet aircraft were investigated. Using the IAD data set, the jet SOTR directivity data were grouped according to a variety of different parameters, and average SOTR directivity adjustments were developed for each of the groups. The SOTR directivity data for turboprop aircraft were not further subdivided, because of the small number of turboprop departures measured at IAD.

One approach was to group aircraft according to engine separation distance, working off the assumption that engine location was dominant influence on the SOTR directivity. It was anticipated that aircraft with similar engine mounting configurations (i.e., fuselage-mount and wing-mount) might have similar directivity patterns. This analysis utilized t-tests to determine if the measured data at each angular offset were similar from aircraft to aircraft. For example, the mean correction at 120 degrees for the A319 was compared to the mean correction for the A320 at 120 degrees. An independent sample t-test was performed to determine if the difference between these two means was statistically significant (i.e., ≠0) at the 0.05 level. These tests answered the question: "Is there at least a 95% probability that the means of these two data sets are statistically similar?" This test was performed at each offset angle for all possible aircraft combinations.

Table 8 shows the number of angular offset points (out of a total of 11) where the average adjustments are similar for each combination of aircraft. Those aircraft combinations that had 10 or 11 similar angles are highlighted in yellow. Those with 9 similar angles are highlighted in a lighter yellow.

Table 8. No. of Angular Offset Points which Exhibit Similarity between Aircraft

	AIR1845	CL600	DC9	B717	B737	A319	A320	B757	B767	A340	A330	B777	B747	Prop
AIR8145		2	2	11	7	8	1	5	1	4	3	2	2	
CL600	2		3	11	3	4	2	2	1	3	1	0	1	
DC9	2	3		10	4	1	2	2	1	3	2	2	3	
B717	11	11	10		10	11	10	10	5	7	4	4	5	
B737	7	3	4	10		11	11	10	3	9	3	1	3	
A319	8	4	1	11	11		10	11	3	8	2	3	3	
A320	1	2	2	10	11	10		11	3	8	2	1	2	
B757	5	2	2	10	10	11	11		9	11	7	4	3	
B767	1	1	1	5	3	3	3	9		11	7	5	8	
A340	4	3	3	7	9	8	8	11	11		9	8	9	
A330	3	1	2	4	3	2	2	7	7	9		11	10	
B777	2	0	2	4	1	3	1	4	5	8	11		11	
B747	2	1	3	5	3	3	2	3	8	9	10	11		
Turbo-Prop	4	5	2	11	9	10	8	10	6	10	4	2	2	

From this table, five groups of aircraft with statistically similar directivity patterns emerge:

1. Fuselage-mounted aircraft - DC9, B717, and CL600
2. Small wing-mounted aircraft - B737, A319, A320 and B757
3. Intermediate wing-mounted aircraft - B767
4. Large wing-mounted aircraft - A330, B777, A340 and B747
5. Turbo-propeller aircraft

As anticipated, aircraft with similar engine mounting configurations exhibit statistically similar directivity patterns. Group 1 consists of aircraft with fuselage mounted-engines. Groups 2, 3 and 4 all consist of aircraft with wing-mounted engines, but the size of the aircraft seems to be the contributing factor between the different groups. Although turbo-propeller aircraft exhibit some similarity to many of the other aircraft, they were considered separately because the general shape of the directivity pattern was not visually similar to any other (i.e., the directivity pattern for the turbo-propeller aircraft does not exhibit a "dip" at 180 degrees). Upon examination of these groupings, the importance of horizontal separation distance between engines was explored as a means to quantify the differences between the above groups. Table 9 presents the engine separation distances for each aircraft type.

Table 9. Horizontal Separation Distances between Engines

Group	Aircraft Type	Approx. Distance Between Outer-Most Engines (ft)
1	CL-600-2B19	13.9
1	DC-9-40	18.4
1	DC-9-30	19.7
1	DC-9-50	20
1	B717-200	21
1	DC-9-82	21
2	B737-500	38
2	B737-700	38
2	B737-800	38
2	B737-300	42
2	A319-100	42.1
2	A320-200	42.3
2	B757-200	48
3	B767-300	56
4	B777-200	64
4	A330-200	68
4	A330-300	68
4	A340-300	66.0 (inner), 134.0 (outer)
4	B747-400	80.0 (inner), 136.0 (outer)

The table shows that the same groupings exist for both aircraft directivity pattern and engine separation distance. The groupings can be described by separation distance as follows:

< 25 ft	- Group 1 type pattern
35-50 ft	- Group 2 type pattern
50-60 ft	- Group 3 type pattern
>60 ft	- Group 4 type pattern

The majority of the aircraft fleet (73%) fell into Group 2. As expected, this resulted in a SOTR directivity adjustment curve very similar to the proposed jet SOTR directivity adjustment from Section 5.4 (within 1.5 dB). While the SOTR adjustments for Groups 1, 3 and 4 all differed from the proposed jet SOTR adjustment (up to 5 dB), the aircraft in these groups represent a small percentage of the fleet (12%, 5% and 10%, respectively).

Another approach was to group the data according various engine parameters. Average thrust level, fan/engine diameter, by-pass ratio, nacelle geometry (short or long exhaust duct), and peak frequency based on a Strouhal relationship (i.e., peak frequency ~ jet velocity/jet diameter) were all considered. When comparing the directivities across different aircraft types, the aircraft with a peak frequency above 120 Hz had a prominent notch or drop-off in the noise levels directly behind the aircraft. This same relationship was also seen in aircraft with average thrust values below 35,000 lb and aircraft with fan/engine diameters smaller than 2 meters. This effect was independent of engine mounting location, by-pass ratio and nacelle geometry. Therefore the jet aircraft were sorted into two different categories: jets with a prominent notch (Group 5) and jets with an absent notch (Group 6). The majority of the aircraft fleet fell into Group 5, which resulted in a SOTR directivity adjustment curve that was very similar to the proposed jet SOTR directivity adjustment from Section 5.4 (within 1.0 dB). While the SOTR adjustments for Group 6 differed from the proposed jet SOTR adjustment (up to 5 dB), it represented a small percentage of the fleet (14%).

The additional SOTR directivity adjustment investigations led to two significant observations. First, the proposed jet SOTR directivity adjustment from Section 5.4 is a good representation for the aircraft comprising the majority of the current fleet. Second, depending on data availability, it may be worthwhile to investigate additional directivity curves or even aircraft-specific SOTR directivity curves.

5.6 Case Studies

5.6.1 Single Event Analysis

The effects of the current and proposed update to the SOTR directivity adjustment (from Sections 1.1 and 5.4, respectively) were assessed on a single event basis in a prerelease, research version of AEDT (AEDT-alpha). Individual departures for four jet aircraft (Boeing 737, Boeing 777, Airbus A320 and Airbus A330) and two turboprop aircraft (Beech 1900D and Saab SF340) from the IAD fleet were modeled in AEDT-alpha, and the sound exposure level (L_{AE}) results were analyzed. Because the behind SOTR adjustments are independent of aircraft type (other than the distinction between jet and turboprop engines), these six aircraft were chosen as representative aircraft, in order to illustrate the typical effect the proposed changes to the SOTR directivity adjustment.

The L_{AE} modeled results were computed at six different locations behind the start of take-off roll segment for each aircraft in AEDT (see Tables 10 and 11). These analysis locations corresponded to measurement locations from the IAD study at different azimuth angles relative to the departure flight track. The differences between the current and proposed SOTR adjustments are presented in Table 12. These differences are equivalent to the expected difference between the SOTR algorithms presented in Equations 1 through 4.

Table 10. Modeled Noise Levels behind Take-off roll Using Current SOTR Adjustment

Metric	Aircraft	Mic (Angle Relative to Take-off Roll)					
		V1 (135°)	V2 (155°)	V3 (180°)	V4 (210°)	V5 (240°)	V6 (270°)
L_{AE} (dB)	Boeing 737	98.1	97.7	89.6	98.2	97.2	96.0
	Boeing 777	106.1	105.9	97.7	106.2	105.1	103.5
	Airbus A320	98.9	98.6	90.6	99.0	97.9	96.7
	Airbus A330	103.0	102.7	94.6	103.1	102.0	100.9
	Beech 1900D	82.9	82.7	74.5	82.9	82.0	81.0
	Saab SF340	91.4	91.0	82.7	91.3	90.6	89.7

Table 11. Modeled Noise Levels behind Take-off roll Using Proposed SOTR Adjustment

Metric	Aircraft	Mic (Angle Relative to Take-off Roll)					
		V1 (135°)	V2 (155°)	V3 (180°)	V4 (210°)	V5 (240°)	V6 (270°)
L_{AE} (dB)	Boeing 737	97.0	94.4	89.1	95.6	95.9	95.5
	Boeing 777	105.0	102.5	97.1	103.7	103.8	103.1
	Airbus A320	97.8	95.3	90.0	96.5	96.6	96.2
	Airbus A330	101.9	99.4	94.0	100.5	100.8	100.4
	Beech 1900D	82.5	77.7	73.4	79.4	80.8	80.4
	Saab SF340	90.8	85.9	81.7	87.6	89.5	89.3

Table 12. Difference in Modeled Noise Levels for Two Different SOTR Implementations (Current – Proposed)

Metric	Aircraft	Mic (Angle Relative to Take-off Roll)					
		V1 (135°)	V2 (155°)	V3 (180°)	V4 (210°)	V5 (240°)	V6 (270°)
ΔL_{AE} (dB)	Boeing 737	1.1	3.3	0.6	2.6	1.3	0.5
	Boeing 777	1.1	3.4	0.6	2.5	1.3	0.4
	Airbus A320	1.1	3.3	0.6	2.6	1.3	0.5
	Airbus A330	1.1	3.4	0.6	2.5	1.3	0.5
	Beech 1900D	0.4	5.0	1.2	3.5	1.2	0.6
	Saab SF340	0.5	5.0	1.0	3.7	1.1	0.4

Table 12 verifies that the different SOTR adjustments were implemented consistently across all six aircraft analyzed.

These single events were also run in a research version of INM 7.0a. The SOTR adjustment implementation was confirmed to be consistent across INM and AEDT. Since AEDT-alpha does not support the capability to generate noise contours, noise contours were generated in the research version of INM 7.0a.

L_{AE} contours for the Boeing 737 jet aircraft were plotted in Figure 47, to illustrate the typical differences in sound level due using both the current and proposed SOTR directivity adjustment. Similar results were observed for Boeing 777, Airbus A320 and Airbus A330 jet aircraft.

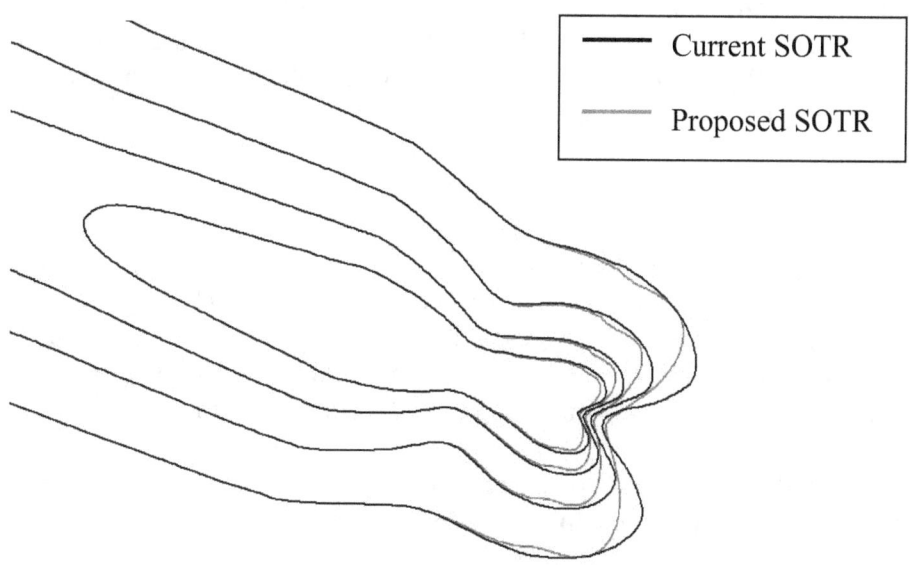

Figure 47. L_{AE} Contours for the Boeing 737 Jet Aircraft with the Current and Proposed behind Start of Take-off Roll Directivity Adjustments for Jet (70 to 85 dB L_{AE} in 5 dB increments)

L_{AE} contours for the Beech 1900D turboprop aircraft were plotted in Figure 48, to illustrate the typical differences in sound level due using both the current and proposed SOTR directivity adjustment. Similar results were observed for the Saab SF340 turboprop aircraft.

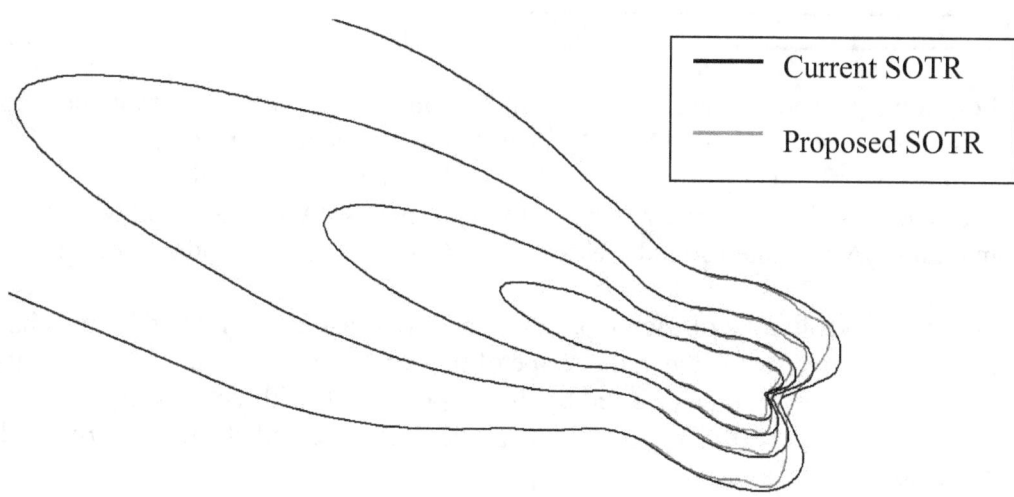

Figure 48. L_{AE} Contours for the Beech 1900D Turboprop Aircraft with the Current and Proposed behind Start of Take-off Roll Directivity Adjustments for Turboprops (70 to 85 dB L_{AE} in 5 dB increments)

For all six of the aircraft, the area for each of the 70 to 85 dB L_{AE} contours decreased by 0.5-2.0% with the proposed SOTR directivity adjustment when compared to the result with the current SOTR adjustment. This decrease in contour area was limited to the area behind the start of take-off roll, as expected.

5.6.2 Airport Analysis

The effects of the proposed change to the behind start of take-off roll adjustment in AEDT and INM were investigated on a study-wide scale. Three international airports (A1, A2, and A3) were modeled in the research version of INM 7.0a with both the current and the proposed SOTR directivity adjustment implementations. All three airports have diverse fleet mixes with both jet and turboprop aircraft.

The differences in day night sound pressure level (L_{DN}) between the current and the proposed SOTR directivity implementation at three different airports are presented in Table 13.

Table 13. Difference in Modeled Noise Levels for Two Different SOTR Implementations (Current – Proposed) at Three Different Airports

Airport	Level (dB L_{DN})	Current SOTR (sq. mi.)	Proposed SOTR (sq. mi.)	Difference (Current – Proposed, %)
A1	55	84.517	83.948	-0.67%
A1	65	14.448	14.288	-1.11%
A2	55	70.449	70.02	-0.61%
A2	65	12.815	12.648	-1.30%
A3	55	92.991	92.486	-0.54%
A3	65	16.188	16.1	-0.54%

For all three airports, the area for each of the 55 and 65 dB L_{DN} contours decreased by 0.5-1.5% with the proposed SOTR directivity adjustment when compared to the result with the current SOTR adjustment. This difference became smaller for quieter noise contours, which were farther away from the runways. A similar trend was observed for maximum A-weighted sound pressure level (L_{ASmx}) contours at the three airports.

While the overall area of each L_{DN} contour is not drastically affected by the change of the SOTR directivity adjustment, the shape of the contour can be affected. Contours near departure runway will be affected by the proposed SOTR directivity adjustment, as illustrated in Figure 49. This would be particularly relevant at smaller airports dominated by turboprop operations.

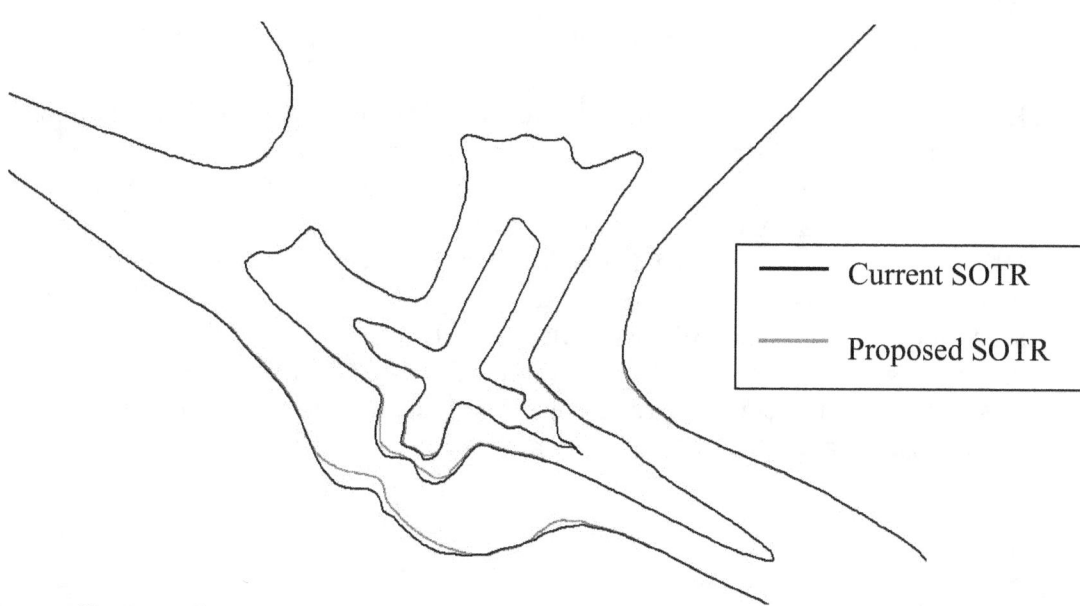

Figure 49. L_{DN} Contours at Airport A2 with the Current and Proposed behind Start of Take-off Roll Directivity Adjustments (55 to 75 dB L_{DN} in 10 dB increments)

6 CONCLUSIONS AND RECOMMENDATIONS

Based on the analysis of the data collected at IAD, it is concluded that the directivity patterns of modern aircraft appear to have only marginal similarities with the existing AIR-1845 directivity pattern. A change in the behind start of take-off roll directivity adjustment implementation to the proposed implementation would benefit AEDT and INM. The proposed adjustments would better represent the current aircraft fleet, and would allow for the directivity associated with jet and turboprop aircraft to be represented separately in the models. While the effects of the proposed SOTR directivity adjustments may be negligible on some full-airport studies, it will allow for a more accurate representation of aircraft departure noise from the current fleet in communities and other noise sensitive areas located near airport runways.

7 FUTURE RESEARCH

While the proposed changes to the SOTR directivity adjustment implementation will better represent the departure noise generated by the current fleet in AEDT and INM, several steps can be taken to further advance this area of research. First, an additional sensitivity analysis could be done for individual aircraft types, to identify (a) if some jet aircraft are better represented by the current SOTR adjustment instead of the proposed SOTR adjustment, and (b) if certain jet aircraft are better represented by aircraft-specific SOTR directivity adjustments. Second, the IAD data set could be reprocessed; in order to establish a set of criteria for determining under what circumstances should additional SOTR directivity adjustments be used. This could include, but is not limited to, the further assessment of the following parameters: engine separation distance; nacelle geometry; engine or fan diameter; by-pass ratio; peak frequency; number of engines, and average thrust level. Third, additional SOTR directivity data could be measured to supplement the Dulles data. This could be especially useful for turboprop and smaller/low thrust jet aircraft, which were only sparsely represented in the IAD data.

REFERENCES

1. Society of Automotive Engineers, Committee A-21, Aircraft Noise, "Procedure for the Calculation of Airplane Noise in the Vicinity of Airports," Aerospace Information Report No. 1845, Society of Automotive Engineers, Inc., Warrendale, Pennsylvania, January 1986.

2. Eldred, K M., Miller, R.L., "Analysis of Selected Topics in the Methodology of the Integrated Noise Model," BBN Report No. 4413, Bolt Beranek and Newman, Cambridge, Massachusetts, November 1980.

3. He, Hua., et. al., Integrated Noise Model (INM) Version 7.0 User's Guide, Report No. FAA-AEE-07-04, FAA, Office of Environment and Energy Washington, D.C., April 2007.

4. Boeker, E.R., et. al., Integrated Noise Model (INM) Version 7.0 Technical Manual, Report No. FAA-AEE-08-01, FAA, Office of Environment and Energy Washington, D.C., January 2008.

5. Gutierrez, R.M., et. al, "Low-Frequency Commercial Aircraft Noise Characterization During Start of Take-Off Roll and Thrust Reverser," Pennsylvania State University, State College, Pennsylvania, 2005.

6. http://www.epa.gov/ttn/scram, September 28, 2005.

7. Federal Aviation Administration (FAA), System for assessing Aviation's Global Emissions (SAGE), Version 1.5, Technical Manual, Report No. FAA-AEE-2005-01, FAA, Office of Environment and Energy, Washington, D.C., September 2005.

8. Federal Aviation Administration (FAA), Volpe Center Acoustics Facility Time-Space-Position-Information System Differential Global Positioning System User's Guide, John A. Volpe National Transportation Systems Center, Cambridge, Massachusetts, July 2001.

9. http://162.58.35.241/acdatabase/NNum_inquiry.asp, September 28, 2005.

10. Statistica Version 7.0, StatSoft, Inc., Tulsa, Oklahoma, 2004.

APPENDIX A: SUMMARY OF INSTRUMENTATION LOCATIONS

Table 14. Summary of Instrument Locations

Instrumentation	ID	Angle (deg)	Distance	Local Coordinates		dGPS Actual Coordinates		Data Type
				X (ft)	Y (ft)	Lat (deg)	Long (deg)	
NASA Acoustic Van 1 mic	N1	120	900	-450	-779	38.93541308	-77.45424897	Acoustic
NASA Acoustic Van 1 mic	N2	135	900	-636	-636	38.93486683	-77.45381874	Acoustic
NASA Acoustic Van 1 mic	N2A	140	900	-689	-579	38.93466755	-77.45370824	Acoustic
NASA Acoustic Van 1 mic	N3A	155	900	-816	-380	38.93403338	-77.45354325	Acoustic
NASA Acoustic Van 1 mic	N4	165	900	-869	-233	38.93357914	-77.45356191	Acoustic
NASA Acoustic Van 1 mic	N5	180	900	-900	0	38.93297614	-77.45373692	Acoustic
NASA Acoustic Van 1 mic	N6	195	900	-869	233	38.93240412	-77.45412636	Acoustic
NASA Acoustic Van 1 mic	N7	210	900	-779	450	38.93193052	-77.45469144	Acoustic
NASA Acoustic Van 1 mic	N8	225	900	-636	636	38.93159738	-77.45539039	Acoustic
Volpe Acoustic System	V1	240	900	-450	779	38.93141035	-77.45618147	Acoustic/Wind Speed/Wind Direction
Volpe Acoustic System	V2	255	900	-233	869	38.93139111	-77.45701438	Acoustic/Wind Speed/Wind Direction
Volpe Acoustic System	V3	270	900	0	900	38.93153509	-77.45782548	Acoustic/Wind Speed/Wind Direction
Volpe Acoustic System	V4	285	900	233	869	38.93183678	-77.45854469	Acoustic/Wind Speed/Wind Direction
NASA Acoustic Van 1 mic	N9	180	1350	-1350	0	38.93253624	-77.45225366	Acoustic
Volpe Acoustic System	V5	105	1950	-505	-1884	38.93819808	-77.45270117	Acoustic/Wind Speed/Wind Direction
Volpe Acoustic System	V6	120	1950	-975	-1689	38.93722718	-77.45140391	Acoustic/Wind Speed/Wind Direction
NASA Acoustic Van 2 mic	N10	135	1950	-1379	-1379	38.93605635	-77.45044372	Acoustic
NASA Acoustic Van 2 mic	N10A	140	1950	-1494	-1253	38.93561349	-77.45022118	Acoustic
NASA Acoustic Van 2 mic	N11A	155	1950	-1767	-824	38.93424189	-77.44985674	Acoustic
NASA Acoustic Van 2 mic	N12	165	1950	-1884	-505	38.93334105	-77.44985950	Acoustic
NASA Acoustic Van 2 mic	N13	180	1950	-1950	0	38.93195737	-77.45028411	Acoustic
NASA Acoustic Van 2 mic	N14	195	1950	-1884	505	38.93072243	-77.45112207	Acoustic
NASA Acoustic Van 2 mic	N15A	140	2500	-1915	-1607	38.93610901	-77.44840779	Acoustic
NASA Acoustic Van 2 mic	N16	180	2500	-2500	0	38.93142382	-77.44847377	Acoustic
NASA Acoustic Van 2 mic	N17A	140	3500	-2681	-2250	38.93701187	-77.44508845	Acoustic
Volpe Acoustic System	V7	180	3500	-3500	0	38.93045647	-77.44518064	Acoustic/Wind Speed/Wind Direction
Volpe Acoustic System	V8	180	4400	-4400	0	38.92958475	-77.44222587	Acoustic/Wind Speed/Wind Direction
Video Tracking System 1	TRACK1	NA	1213	449.5	1126.2	38.93138737	-77.45957142	Aircraft Position
Video Tracking System 2	TRACK2	NA	3133	2617.2	-1722.9	38.94080657	-77.46317287	Aircraft Position
Meteorological Station	MET	NA	292	-250	-150	38.93212000	-77.45447000	Wind Speed/Wind Direction/Temp/RH/BP
Tail Logging System (Primary)	LOG1	NA	292	-250	-150	38.93138737	-77.45547000	Aircraft Tail Numbers
Tail Logging System (Secondary)	LOG2	NA	922	200	900	38.93138737	-77.45957142	Aircraft Tail Numbers
Origin	Origin	NA	0	0	0	38.93384721	-77.45670386	NA

APPENDIX B: TEST DIRECTOR LOG SHEET

Table 15. Test Director Log Sheet

Volpe Center Acoustics Facility

Test Director Log

Test Director:		Date:		
Location:		Start Time:		Page __ of __
		End Time:		

Event ID	Event Start Time	Event End Time	Avg. Wind Speed (mph)	*Idling Aircraft Nearby? (Y/N)*	*Event Good? (Y/N)*	Comments

APPENDIX C: TAIL NUMBER LOG SHEET

Table 16. Tail Number Log Sheet

Aircraft Tail Number Log

Volpe Center Acoustics Facility

Logger:	Date:	Page ___ of ___
Location:	Start Time:	
	End Time:	

Event Start Time	Event End Time	Event ID	Tail Number	SOTR Type		Airline	Body Type				No. of Engines	Engine Config		Comments
				Rolling	Static		Regional Jets	Turbo Props	Narrow Body	Wide Body		Tail Mounted	Wing Mounted	

APPENDIX D: VOLPE NOISELOGGER SYSTEM SETUP

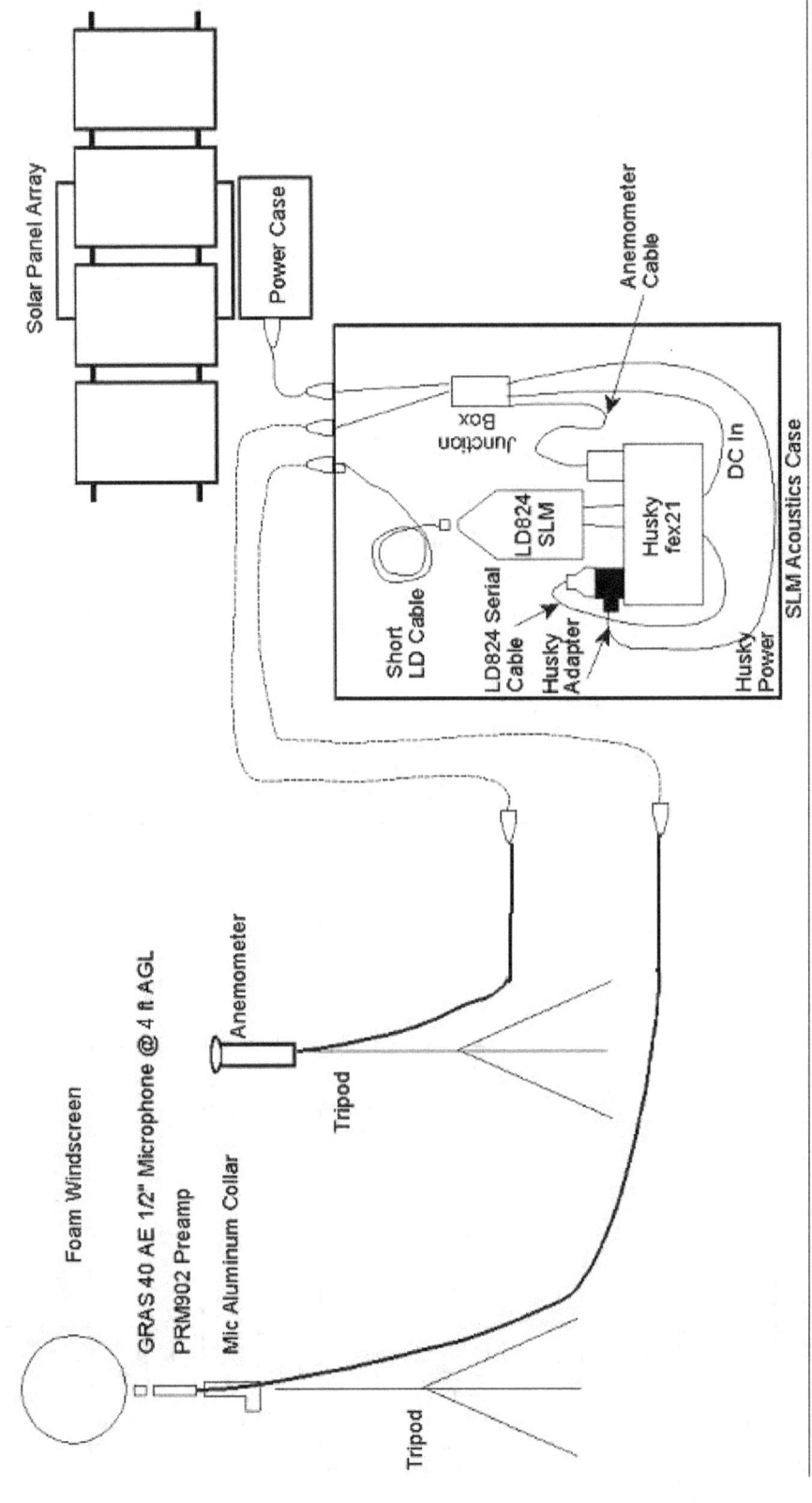

Figure 50. Volpe NoiseLogger™ System Setup

APPENDIX E: DIRECTIVITY PATTERNS AT 1950 FT

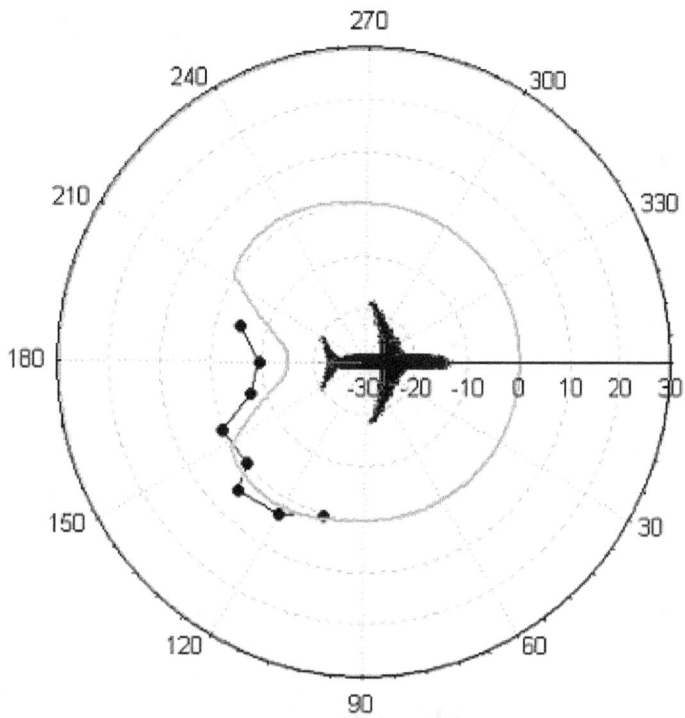

Figure 51. Airbus A310 Directivity Pattern at 1950 ft (1 Event)

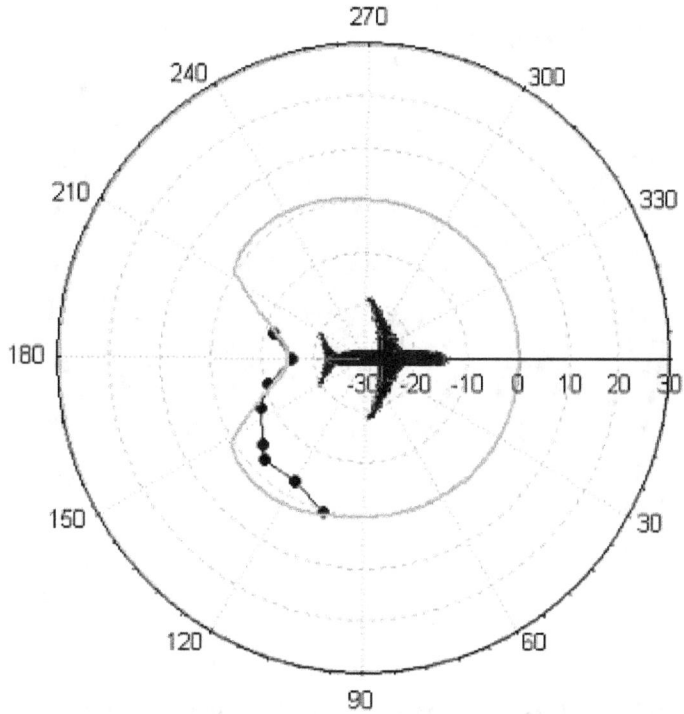

Figure 52. Airbus A319 Directivity Pattern at 1950 ft (16 Events)

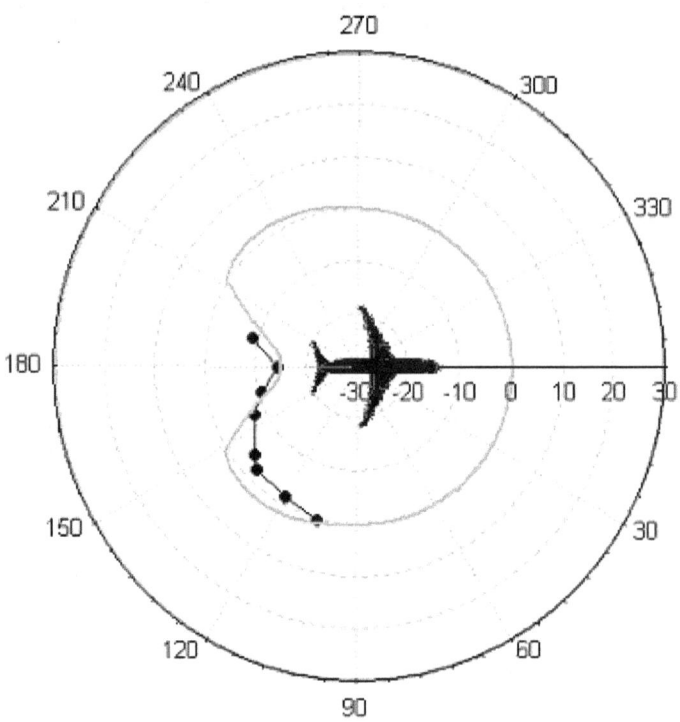

Figure 53. Airbus A320 Directivity Pattern at 1950 ft (48 Events)

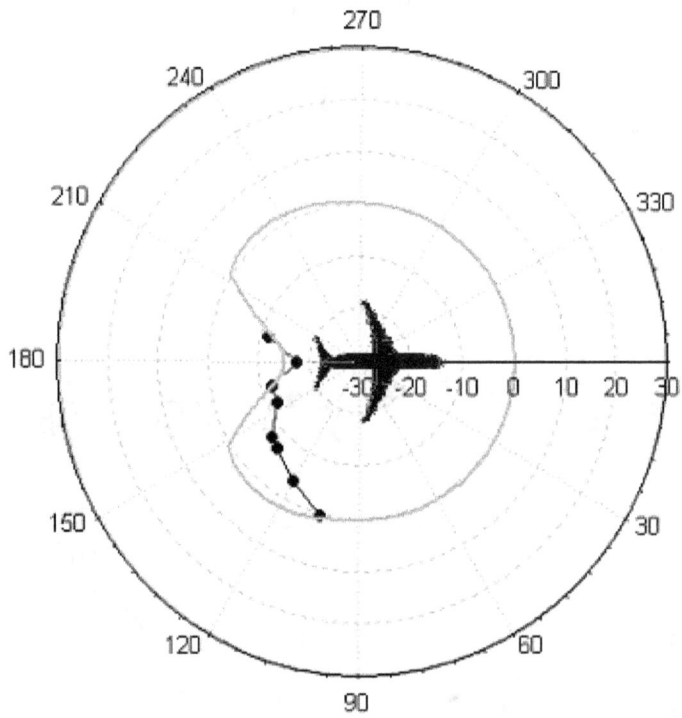

Figure 54. Airbus A330 Directivity Pattern at 1950 ft (4 Events)

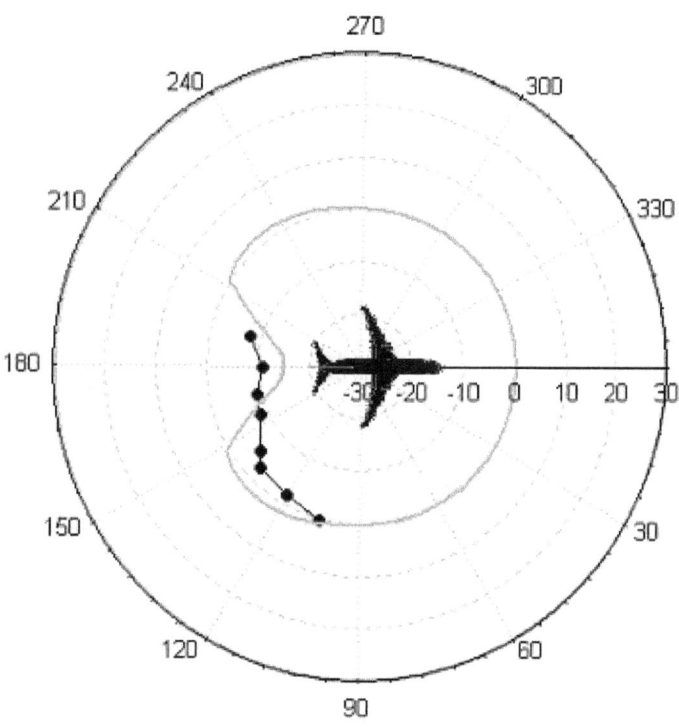

Figure 55. Airbus A340 Directivity Pattern at 1950 ft (5 Events)

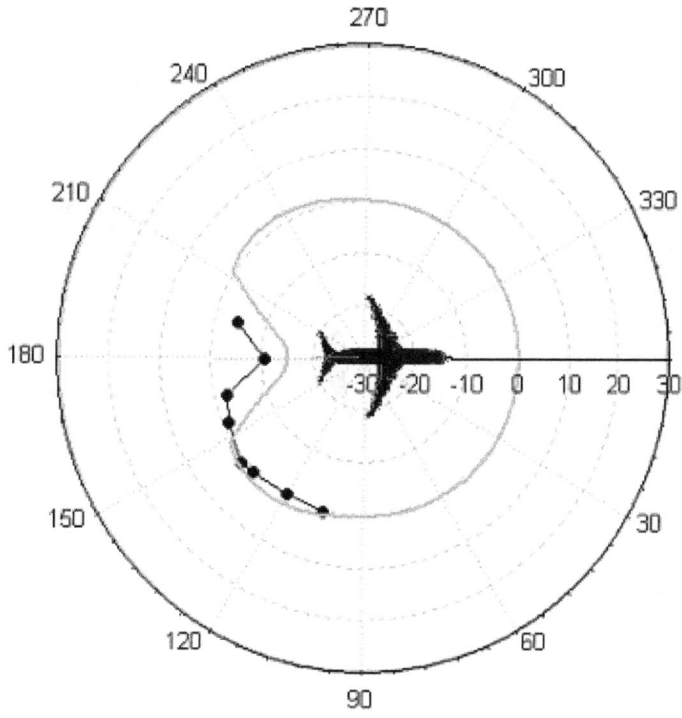

Figure 56. Boeing 717 Directivity Pattern at 1950 ft (3 Events)

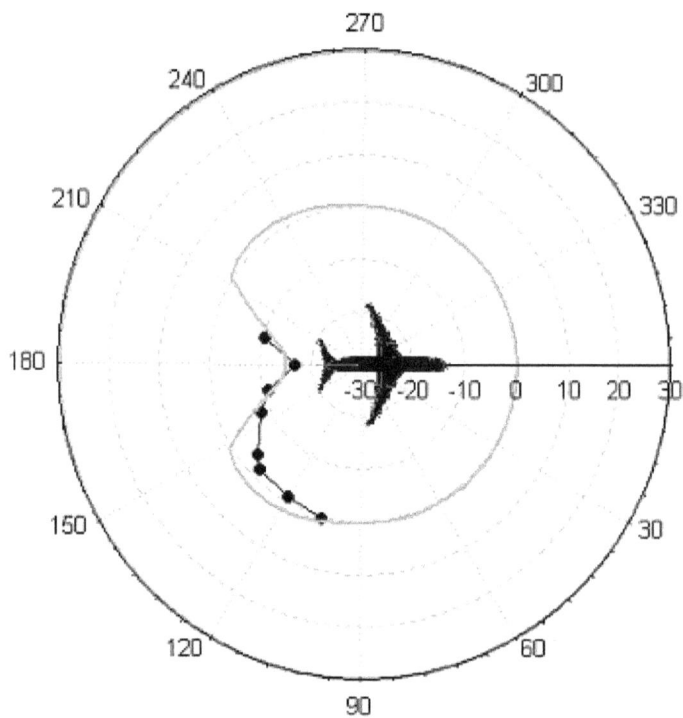

Figure 57. Boeing 737 Directivity Pattern at 1950 ft (20 Events)

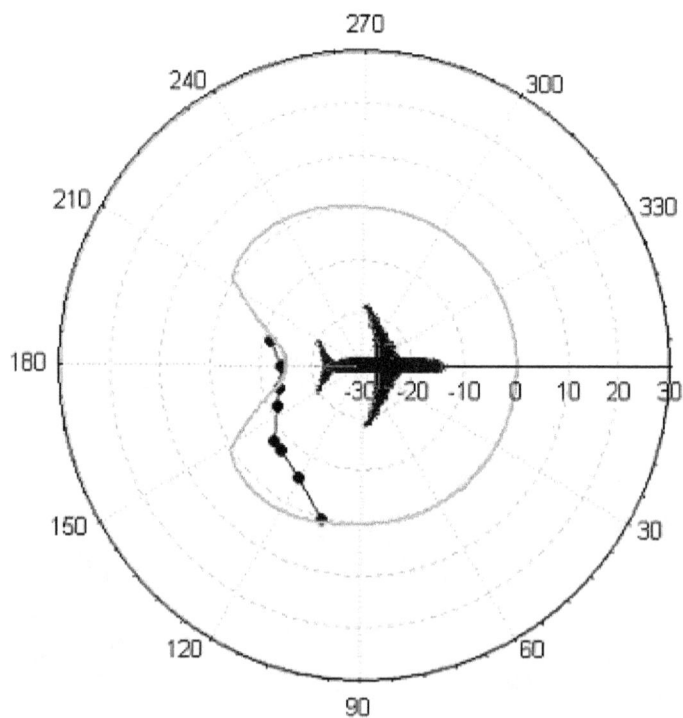

Figure 58. Boeing 747 Directivity Pattern at 1950 ft (7 Events)

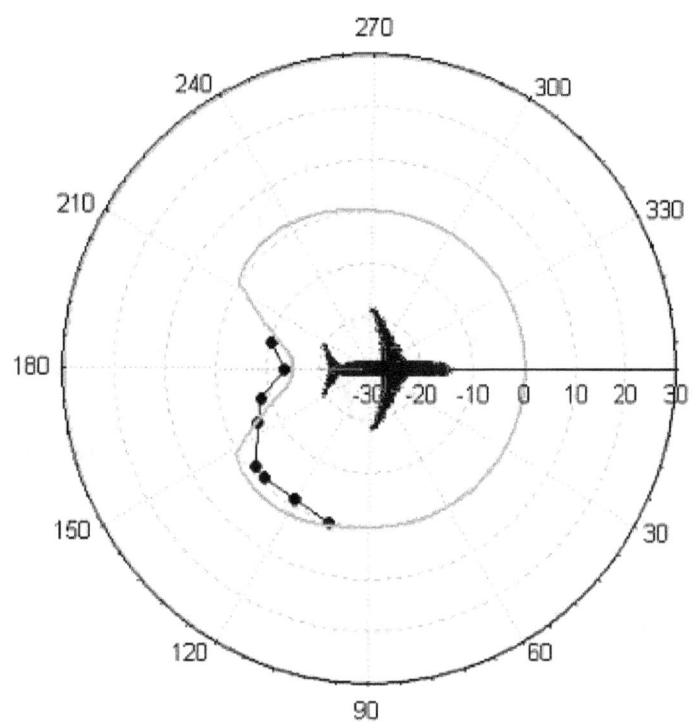

Figure 59. Boeing 757 Directivity Pattern at 1950 ft (12 Events)

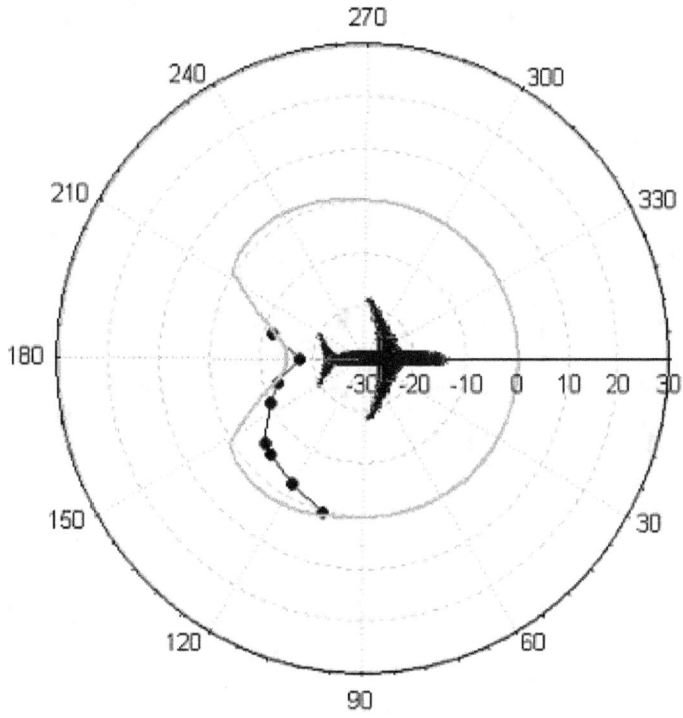

Figure 60. Boeing 767 Directivity Pattern at 1950 ft (27 Events)

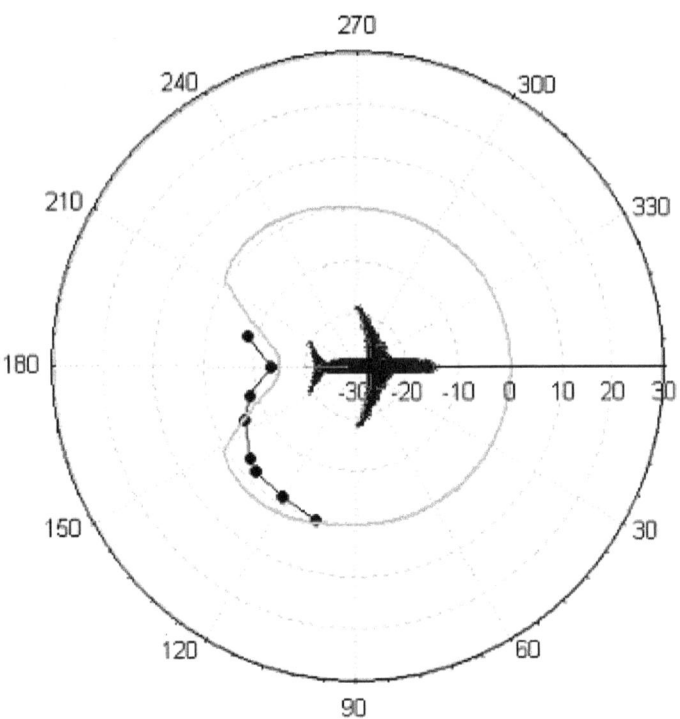

Figure 61. Boeing 777 Directivity Pattern at 1950 ft (12 Events)

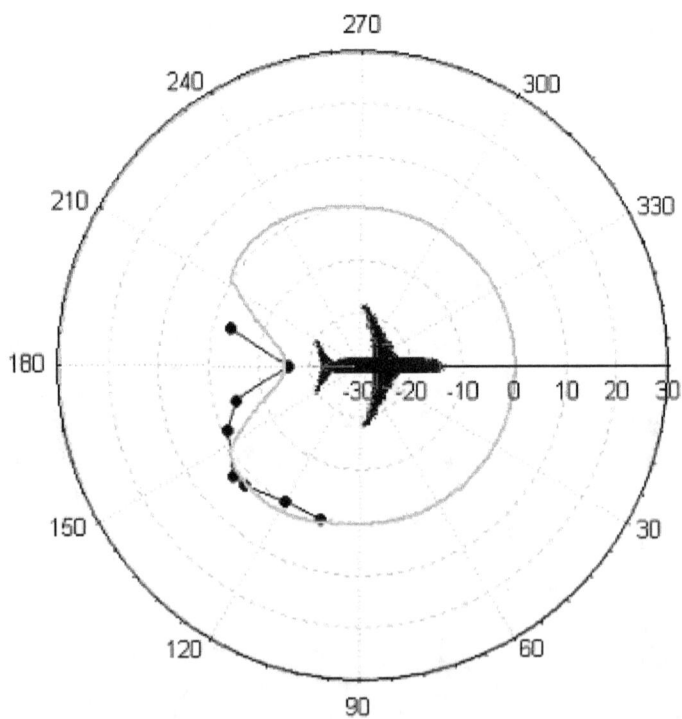

Figure 62. McDonnell Douglas DC9 Directivity Pattern at 1950 ft (23 Events)

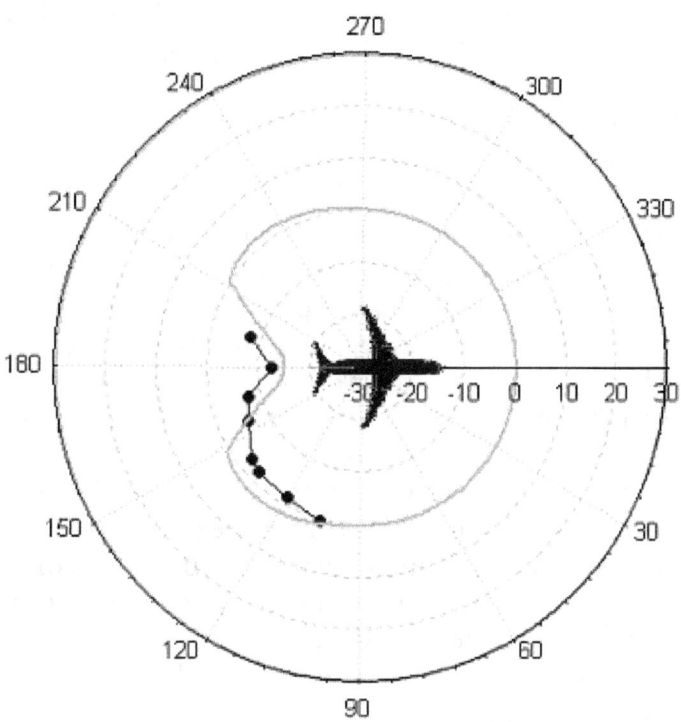

Figure 63. Bombardier CL600 Directivity Pattern at 1950 ft (104 Events)

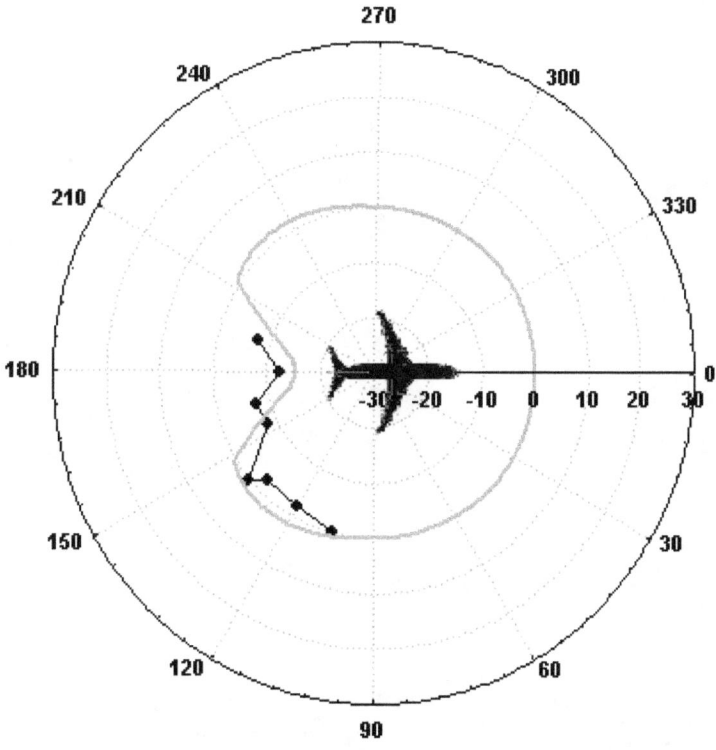

Figure 64. Turbo-Propeller Aircraft Directivity Pattern at 1950 ft (14 Events)

APPENDIX F: AIRCRAFT FLEET PERCENTAGE FOR DIRECTIVTY ADJUSTMENT DERIVATION

Two sets of -weighted SOTR directivity adjustments were developed; one for jet aircraft, and the other for turboprop aircraft.

For jet aircraft, a weighted-regression was used to compute the SOTR directivity adjustment. The weighting factor was chosen to modify the regression such that aircraft were represented proportionate to their relative operational presence in the 2005 fleet[*]. The number of 2005 operations and corresponding percentage of each of the jet aircraft measured is presented in Table 17. As shown in the rightmost column, the aircraft measured during this study represent 85% of the total operations in 2005.

The weighting factor applied to the average noise level (in decibels) for each aircraft at each azimuth angle behind SOTR reflects the percentage of the *measured jet fleet*. As the measured jet fleet does not include all aircraft in the total fleet (15% of aircraft were not measured and 23% were propeller aircraft), weighting factors were adjusted to reflect only the portion of jet aircraft included in the measurement data. For example, the Airbus A319 represents only 5% of total operations in 2005, but 9% of operations when only those corresponding to aircraft in the measured jet fleet are considered. Thus, weighting factor of 9 was applied to the data for the Airbus A319[†].

---

[*] AEDT Common Operations Database (COD) for the FAA Destination 2025 environmental inventory of the calendar year 2005.
[†] Weighting factors were entered as integers, and the total of all weighting factors equaled 100.

Table 17. Jet Aircraft Fleet Percentages for Directivity Adjustment Derivation

Aircraft	Number of Operations	% of Measured Jet Fleet	% of Total 2005 Operations
Airbus A319	1423985	9%	5%
Airbus A320	2372032	14%	9%
Airbus A330	322430	2%	1%
Airbus A340	190224	1%	1%
Boeing 717	333272	2%	1%
Boeing 737	7088418	43%	27%
Boeing 747	595303	4%	2%
Boeing 757	1162127	7%	4%
Boeing 767	823779	5%	3%
Boeing 777	453399	3%	2%
McDonnell Douglas DC-9	339756	2%	1%
Bombardier CL-600	1353937	8%	5%
Propeller Aircraft	*6040263*	*N/A*	*23%*
Not Measured	*3834201*	*N/A*	*15%*
Total:	**23274963**	**100%**	**100%**

For turboprop aircraft, a more standard regression fitting process was used to develop the SOTR directivity adjustment. The regression was computed using all data from the events measured. No weighting factor was applied, as the measured data set was smaller (14 events) and could not adequately represent the majority of the turboprop fleet. As a result, the Saab SF340 is the predominant aircraft represented in this regression (10 of 14 events), while the Cessna 425 (1 of 14 events), Rockwell 690B (2 of 14 events) and Beech 1900D (1 of 14 events) are represented to a lesser degree.

www.ingramcontent.com/pod-product-compliance
Lightning Source LLC
Chambersburg PA
CBHW081834170526
45167CB00007B/2800